NUCLEAR P⚛WER DEVELOPMENT

Prospects in the 1990s

Stanley M. Nealey

 **BATTELLE PRESS**

Columbus • Richland

A portion of the funding for this research was provided by the U.S. Department of Energy. However, the views and findings are solely those of the author and do not necessarily represent the policy of the agency. The agency encourages wide dissemination of the technical information contained herein, with due respect for the Publisher's rights regarding the complete volume.

Library of Congress Cataloging-in-Publication Data

Nealey, Stanley M., 1934–
   Nuclear power development: prospects in the 1990s/Stanley M. Nealey.
   p.   cm.
   Includes bibliographical references.
   ISBN 0-935470-53-0:$18.95
   1.   Nuclear energy—United States. I. Title.
TK9023.N43   1989
333.792'415'0973—dc20

89-38237
CIP

Battelle Press
505 King Avenue
Columbus Ohio 43201-2693
614-424-6393; 1-800-451-3543
FAX: 614-424-5263

PREFACE

*A*lthough nuclear power currently accounts
for almost 20 percent of the electricity
generated in the United States, with over one hundred
reactors in operation, the rate of nuclear power
development has been declining since the early 1970s
when nuclear plants on order and under
construction began to be cancelled. The objective of
this study is to examine the prospects for a
resumption in growth of nuclear power in the United
States over the next decade. The focus of analysis is
on the likelihood that current efforts in the United
States to develop improved, and safer, nuclear power
reactors will provide a sound technical basis for
improved acceptance of nuclear power and
contribute to a social/political climate more
conducive to a resumption of nuclear power growth.

   The acceptability of nuclear power and
advanced reactors to five social/political sectors in the
U.S. are examined: 1) the financial sector involved in
financing nuclear power plant construction, 2) the

federal nuclear regulatory sector, 3) the national political sector, 4) those in the general public who are knowledgeable about and involved in nuclear power issues, the "involved public," and 5) the much larger body of the general public that is relatively "uninvolved" in the controversy over nuclear power.

Nine reactor technology development features are examined in order to assess their relation to improved financial, regulatory, political and public acceptance. Three technology features were rated as highly important in gaining financial support. These were smaller plant size, shorter and more predictable construction time, and design standardization. Of moderate importance to the financial community were improved operating ease, lower construction cost, and factory construction capability. The potential for inherent safety to win public and political support is valued by the financial community even though most already view current power reactors as safe enough. Operating cost and modular increments to plant size are viewed as less important than the other reactor features analyzed.

An improved and more predictable regulatory climate is generally felt to be necessary if new reactors under development are to be successfully deployed. Observers of the Nuclear Regulatory Commission (NRC) are not confident that significant relief from current regulations is a likely prospect. The NRC appears to welcome new technology development efforts and is cooperating with them. However, the social/political environment in which NRC operates is likely to make major change a slow and uncertain process.

Despite industry calls for the government to exert strong leadership, political support for new reactor development and deployment will probably follow rather than lead public acceptance. Survey data show the views of politicians quite faithfully mirror those of the general public. Since enhanced plant safety appears to be the primary avenue to public acceptance, it follows that reactor technology development ought to garner political support.

Public attitudes toward nuclear power in the United States were assessed by Nealey, Melber, and Rankin (1983). This analysis is updated through 1987 by examination of eight public attitude topics that bear on further nuclear power development. Public support for building nuclear power plants declined from strong majority support prior to Three Mile Island (TMI), to a break-even level by 1982, to four to one opposition in 1987. Support for constructing a plant in one's local area began to decline even before TMI. Public support for proposals to curtail nuclear operations in order to solve safety problems have had majority public support since TMI, yet only about 15-20 percent of the public has supported permanent closure, and every state initiative to close operating plants has been defeated. However, nuclear industry views that nuclear power has been over-regulated are not shared by the public; it continues to favor very tight regulations.

Nuclear power plant safety dominates other issues as judged by survey results. Nuclear power currently has low salience as a social issue to most Americans. However, safety, as the most dramatic nuclear issue, commands attention even among the "uninvolved" public. Despite the fact that about half the public characterize nuclear power plants as "dangerous," few feel personally threatened and a majority feel safety has improved in the past ten years.

Media accounts of the Chernobyl accident were widely followed by the public and a source of widely felt, but not strongly felt, concern. Most recognized that U.S. reactors were different and safer than Soviet reactors, but felt that an accident of similar severity could happen here.

Compared to other electricity generating technologies, nuclear power is least favored by the public, despite public recognition of its relative environmental advantages and a better safety record than coal. However, a large majority of the public continues to feel that nuclear power will play an increasingly important role in supplying electricity in the future. The reluctance with which nuclear power is accepted by most members of the public is evidenced by its image as a "realistic" rather than a "good" choice.

A hard core of opposition to nuclear power constitutes about one in four or five Americans. Ten percent or so of the public remains uncommitted, and the views, positive or negative, held by many members of the public are not strongly held views. Thus, many members of the public are open to the influence of future events and new information.

Nuclear plant safety is highly important to both the involved and uninvolved public as well as the regulatory and political sectors. Whether public confidence in the safety of light water reactors can be won, even given the substantial safety improvements currently under development, is a question of great importance. No definitive answer is possible, but the prospect that the advanced safety reactors could substantially remove concerns about catastrophic accidents gives them special status among the reactor development paths. It is essential for the new reactors to capture favorable media attention if they are to win public acceptance. It is too early to tell whether media attention to global warming will contribute to significant public concern that could improve the chances for nuclear power growth. However, should a scientific consensus develop that fossil fuel burning must be curtailed, and should the current trend to higher electricity demand growth continue, the question of nuclear power growth will receive much greater public and political attention.

If there is to be a resumption of nuclear power development, four conditions must be met: 1) there must be a clear need for more base-load power, 2) the next generation of nuclear plants must present clear life-cycle cost superiority over coal, 3) nuclear plant construction costs must be predictable and lower, and 4) the public must be convinced that a proposed nuclear power plant will be safe. Development and demonstration of the advanced safety reactors stands a reasonable chance of generating the favorable media attention necessary to public and political acceptance.

CONTENTS

TABLES

LIST OF
ABBREVIATIONS

(AE)	Architect Engineering Firm
(AFUDC)	Allowance for Funds Used During Construction
(CWIP)	Construction Work in Progress
(DOE)	U.S. Department of Energy
(EPRI)	Electric Power Research Institute
(FERC)	Federal Energy Regulatory Commission
(HTGRs)	High Temperature Gas Cooled Reactors
(LMRs)	Liquid Metal Reactors
(LOCA)	Loss of Coolant Accident
(LWRs)	Light Water Reactors
(NIMBY)	"Not-in-My-Backyard"
(NRC)	U.S. Nuclear Regulatory Commission
(ORNL)	Oak Ridge National Laboratory
(OTA)	U.S. Office of Technology Assessment
(PIUS)	Process Inherent Ultimately Safe
(PUC)	Public Utility Commission
(TMI)	Three Mile Island
(USCEA)	U.S. Council for Energy Awareness

NUCLEAR P⚛WER DEVELOPMENT

Prospects in the 1990s

INTRODUCTION   1

The rapid growth of nuclear power in the United States as a source of electricity began in the 1960s. By the time the last reactors currently under construction come on line in the early 1990s, the number of reactors with operating licenses will total 115-120 and will likely account for over 20 percent of the electricity generated in the United States. Nuclear power supplies more electricity to the grid than any other source except coal. Yet this growth started to show signs of ending in the early 1970s when utilities began to cancel orders for new nuclear power plants. Between 1972 and 1983 over 100 plants were cancelled (Office of Technology Assessment (OTA), 1984). No new plants have been ordered since 1978, and in 1988 no utility appears even close to ordering a new plant. While public concern over the safety of nuclear power plants is widely believed to be a major factor in the decline of nuclear power, plant cancellations actually began during an era of widespread public support for continued nuclear power plant construction. National public opinion polls showed a margin of nearly two to one in favor of continued nuclear power plant construction during the period 1976-1979 (Nealey, Melber and Rankin, 1983). This margin of support was eroded somewhat after the accident at the Three Mile Island (TMI) Unit 2 in 1979. Yet up until 1982, polls continued to show more of the public in support than in opposition to continued nuclear power plant construction (Nealey et al.; 1983; OTA, 1984). By 1983, national polls that asked the question, ''Do you favor or oppose the construction of more nuclear power plants?,'' showed the public in opposition by about two to one (Schneider, 1986; OTA, 1984). By 1986, opposition had increased to more than four to one (Schneider, 1986).

Support for building a plant locally is a different issue from that of support for nuclear power plant construction in general. There were signs as far back as 1976 that support for building nuclear plants in one's local area was on the decline. By 1978, opposition to local plant construction had grown to the point where it exceeded support (Nealey et al., 1983). However, these data should not be interpreted as evidence that the American public has given up on nuclear power. Survey data reported in Chapter 3 make it clear that a substantial majority of the public in 1988 felt nuclear power will be important in meeting the nation's electricity needs in the years ahead.

While orders for new plants are not an immediate prospect, the contribution of nuclear power to the electricity supply continues to increase as new plants are completed. In 1987, 17.7 percent of the electricity generated in the United States came from nuclear reactors, and eight new plants started up. This brought the total of plants with operating licenses to 109, with 14 more still under construction (U.S. Council for Energy Awareness (USCEA), 1988).

Whether reduced public confidence in the safety of nuclear power played a decisive and direct role in halting nuclear power expansion in the United States is an important question. If it did, then the implication is that regaining public confidence is a necessary condition for the resumption of nuclear power growth. In any case, public support is a valuable component of the mix of technical/social/political factors necessary to nuclear power expansion.

There is little doubt that one major factor in halting nuclear power expansion was a sharp drop in demand growth, beginning in 1973. Demand growth averaged 7.1 percent a year in the period 1960-1972; it droped to 2.6 percent a year in the period 1973-1982 (OTA, 1984). Since 1982 load growth has averaged 3.3 percent a year, rising to 4.5 percent in 1987 (USCEA, 1988). Several other factors in addition to this sharp decline in demand growth have been credited with halting nuclear power expansion. Chief among these are: 1) a loss of confidence by the financial community in the ability of utility and nuclear industry management, 2) increasingly complex and burdensome regulation at both the federal and state level, and 3) erosion of the historic cost advantage that nuclear power has enjoyed over coal as the major option for large-scale, base-load electricity generation in the United States. Nuclear power expansion (as well as expansion of coal-fired generating capacity) may also have been affected by a sharp increase during the period 1973-1975, in the public's belief that conserving energy was preferable to constructing more power plants (Nealey et al., 1983). All of these factors, plus others such as problems with radioactive waste disposal, played some role in the decline of nuclear power.

The objective of this study is to examine these factors and the prospects for a resumption in growth of nuclear power in the United States over the next decade. The focus of analysis will be on the likelihood that current efforts in the United States to develop improved and safer nuclear power reactors will provide a sound technical basis for improved acceptance of nuclear power, and contribute to a social/political climate more conducive to a resumption of nuclear power growth. The acceptability of nuclear power and advanced reactors to five social/political sectors in the U.S. will be examined. Three sectors highly relevant to the prospects for a restart of nuclear power plant construction are the financial sector involved in financing nuclear power plant construction, the federal nuclear regulatory sector, and the national political sector. For this analysis, the general public will be divided into two groups: those who are knowledgeable about and involved in nuclear power issues, the "involved public," and the much larger body of the general public that is relatively "uninvolved" in the controversy over nuclear power.

Briefly, the advanced reactor development programs involve: 1) improved light water reactors (LWRs), 2) advanced design LWRs incorporating various features such as smaller size, greater design simplicity, greater operating ease, and various so-called "passive safety" or "inherently safe" features designed to remove decay heat in the event of a malfunction, and 3) liquid metal reactors (LMRs) and high temperature gas cooled reactors (HTGRs) designed with inherent safety and improved efficiency as major design criteria. These three reactor development paths will be referred to as "refined LWRs," "advanced LWRs," and "advanced safety designs," respectively.

The refined LWRs are essentially ready to deliver. They incorporate the lessons of the past as applied to existing LWR designs. The advanced LWRs are being developed by several reactor vendors, partly in response to a program of the Electric Power Research Institute (EPRI) which had input from utilities and reactor vendors (EPRI, 1982, 1986). EPRI's program developed a set of requirements involving improved safety, reliability, operability, maintainability and constructability. Most vendors have developed designs of the 600 MWe size as well as designs in the 1300-1350 MWe range. They anticipate substantially reduced construction times (three to five years) and expect design approvals from the Nuclear Regulatory Commission (NRC) before the mid-1990s. If these expectations are realized, some of these designs could be ready to construct in five to seven years.

The advanced safety designs are under development with support from the U.S. Department of Energy (DOE). The design goals take the EPRI guidance into account with special emphasis on inherent safety. Improved public acceptance is an explicit objective of this program, as is reduced cost. The DOE's Nuclear Energy Long-Range Facility Utilization Plan states a goal of "developing and demonstrating advanced nuclear power plants by the turn of the century" (Draft, March 1978, p.1). The HTGR designs under development range from 200-550 MWe per reactor and are conceived of as production modules that could be built sequentially at a site, thus enabling better response to growth in demand for electricity. Projecting the availability of these advanced safety designs involves many uncertainties. The treatment they receive by the NRC will surely affect the schedule. The nature and success of demonstrations will also set limitations on their availability. Orders cannot be expected until technical and institutional questions are answered. It appears some designs may be ready to construct in seven to ten years or perhaps a bit longer.

Nine features of reactor technology development have been identified as a framework for the analysis of potential public, political, financial and regulatory acceptance. Since there are thousands of design elements involved in reactor development, each of these nine features involves a number of elements. The nine features are also somewhat interrelated. They were chosen as a convenience to the analysis, rather than as a precise definition of the scope of reactor development.

Plant Size

Most designs involve plants smaller than the LWRs in excess of 1000 MWe constructed in the past decade. Although plant designs have tended to increase in size over the course of nuclear development in the United States, there is a body of opinion that economies of scale have reached, or perhaps exceeded, a point of diminishing return (Fisher, Paik and Schriber, 1986; Energy Information Administration, 1986).

Inherent Safety

Safety improvement has always been a reactor design objective, one that can be approached in a variety of ways. To a greater or lesser extent, reactor development now underway involves special features that allow for the removal of decay heat in the event of a malfunction. Some of these features have been called inherently safe. They involve passive systems that work on thermal/mechanical principles which do not depend on active systems, operator control or availability of operating power. It should be noted that "inherent safety," "passive safety" and "fail safe" are terms that denote a design dimension rather than an absolute. Indeed, several speakers at the American Nuclear Society meeting on the "Safety of next-generation power reactors" (held in Seattle, Washington, May 1988) cautioned against using these terms lest they be interpreted to mean zero risk.

Operating Ease

The fact that the power plant accidents at Brown's Ferry, TMI and Chernobyl all involved substantial human error is in line with the conclusion that human error is the dominant risk for severe accident occurrence in nuclear power plants (Haney and Blackman, 1987). Operating ease incorporates a number of elements, such as greater operating safety margins, improved operator/machine information and control interfaces, fewer operating elements, and the use of computers and automation. The general goal of this feature is to minimize the potential for human error, although easier maintenance and greater resistance to equipment damage are also important goals.

Construction Time

Reducing the construction time of nuclear power plants has become an important objective. While there are some recent exceptions, nuclear power plant construction time more than doubled during the 1970s (OTA, 1984). Many plant construction schedules were deliberately delayed in response to the decline in demand growth. However, a number of other factors also contributed to construction delays, including regulatory requirements to install additional safety equipment in the wake of TMI, labor problems, material shortages, financing problems, and problems of incomplete or faulty design (OTA, 1984). Many of these difficulties are referred to collectively as "management problems," yet some are well within the scope of technology development. Building in rather than adding on safety features is generally cheaper and quicker. Less complex designs are also generally quicker to build than more complex ones. Predictability of construction time is a variable of importance independent of actual construction time. Unexpected delays cause financial and institutional problems. Therefore, one goal of reactor development efforts is to make plant construction more predictable. This reactor feature, therefore, refers to a variety of design elements that affect construction time and the predictability of the course of construction.

Construction Cost

Since capital cost amounts to such a large fraction of the life-cycle cost of nuclear power (estimated as high as 70-80 percent) (Delene, Bowers and Shapiro, 1988; Lennox and Mills, 1988), keeping the cost of construction as low as possible has important effects on the economic viability of nuclear power. Nuclear plant construction costs more than doubled in constant dollars on the basis of kWh generating capacity during the 1970s and have increased further in the 1980s. While some recently completed plants have come in as low as $1,000 per kWh, others have cost $3,000 per kWh or more (OTA, 1984). Based on capital cost data from a small group of recently completed single-unit nuclear plants with the best industry cost experience, a recent report from Oak Ridge National Laboratory (ORNL) estimates capital costs for a 1,100 MWe nuclear plant at $1,690 per kWh (Williams, Delene, Fuller and Bowers, 1987) in 1986 dollars, about 30 percent higher than the cost of a coal plant of the same size. Other industry estimates project even lower costs for nuclear plants. For instance, a study group of the USCEA (1987) predicts nuclear plants could be built for $1,075 per kWh in 1986 dollars, *if* plants were standardized and enjoyed a "stable regulatory environment." Nuclear plant cost estimates appear to be driven by the assumptions on which analyses are based. The point is that capital costs must be lowered for nuclear power to be competitive. The 1984 Office of Technology Assessment Report (OTA, 1984, p.66) asserts, "When the construction cost of a nuclear plant is no more than 20 to 40 percent above that of a coal plant, it can be expected to produce electricity more cheaply, sometimes substantially, over the plant's lifetime." The potential for increased or decreased plant cost appears to be an important feature of reactor technology development.

Operating Cost

As with construction costs, a number of operating cost factors are independent of reactor technology. Fluctuations in fuel cost and some regulatory actions are examples. Operating cost and operating ease are somewhat related. A plant designed for easy maintenance would benefit both features. Operating cost is affected by the basic efficiency of the design, but the percentage of time a plant is available to produce electricity has substantial impact on operating cost. Plants designed and built to suffer little down time are more likely to be competitive.

Modular Increments to Plant Size

One of the penalties of large plants, regardless of economies of scale, is that they have difficulty matching slow demand growth. Plants that could be built in increments of small production units over time could match the demand curve more effectively. It appears some of the HTGR and LMR designs are tailored to do this, although any design that could be built and operate economically in small increments would enjoy this advantage.

Design Standardization

Standardization of nuclear plant design has been a long-term but elusive goal of nuclear power in the United States. The potential for design standardization to make plants easier to build, operate and regulate has long been recognized (OTA, 1981). Institutional rather than technical factors appear responsible for the general lack of standardization in the past. Given the large number of utilities that make purchase decisions in a competitive market supplied by several vendors and many architect engineering firms (AEs), the lure of technical advances and projected cost advantages has mitigated against standardization. Nonetheless, the opportunity remains. Reactor development efforts that can build on past experience may find the institutional climate for standardization to be more favorable than in the past.

Factory Construction

Building plant components at factories for assembly at the plant site is obviously related to both plant size and design standardization and could, in principle, improve quality and reduce cost of construction. Prospects for multiple orders would be required to justify the development of factory production facilities. Plant components would have to be designed in units small enough to transport. Past efforts along this line have made some progress, but the opportunity remains to make much greater use of factory construction techniques.

It is beyond the scope of this study (and the competence of its author) to pass judgment on the technical merits of the three reactor development paths as they relate to the nine nuclear plant features just discussed. The nine nuclear plant features will be used as a rough framework for assessing the prospects for nuclear power growth.

It is not possible to forecast when new orders for nuclear plants of whatever kind will be placed in the United States. A variety of factors independent of reactor technology development will influence the timing of new orders (Denton, 1983). Seven such factors will be discussed. Economic conditions as reflected in the *growth in demand for electric power* and in the *level of interest rates* will set limiting conditions on nuclear power growth regardless of developments in reactor technology. In a climate of very slow growth in demand (one to two percent a year) and in a climate of very high interest rates, nuclear power development is unlikely. Owing to the substantial construction lead times for nuclear plants (even if shorter than in the past) and the fact that the life-cycle costs of nuclear power are heavily weighted in front-end capital costs, slow demand growth and high interest rates combine to substantially increase the cost of nuclear generated electricity. On the other hand, demand growth in the five percent range, or higher, coupled with low interest rates would favor the development of capital-intensive base-load energy supply technologies.

Another significant cost factor is the *rate-setting treatment* which nuclear power receives from state public utility commissions (PUCs). Nuclear power, or any capital intensive electricity producing technology, must be able to recover its investment on a schedule that keeps it competitive with alternatives that feature lower capital costs but higher life-cycle costs, e.g., combustion turbines. If PUCs do not allow a rate structure that provides allowance for funds used during construction (AFUDC) to be recovered by charging rate payers for construction work in progress (CWIP) the prospects for further nuclear development are reduced (OTA, 1984).

Several aspects of current nuclear power, both in the United States and abroad, could substantially influence domestic nuclear power growth, independent of technology development. Chief among these is the *performance of currently operating reactors.* A flood of continuing operational, quality control, and safety problems could prevent the necessary growth in confidence in nuclear management that would permit new nuclear power growth even if new and better technology is available.

It is also clear that another *major nuclear plant accident* would delay further nuclear plant deployment. The accident at TMI in 1979 did not precipitate the wave of plant cancellations, but it likely added to it. While the impact of the Chernobyl accident on public opinion in the United States appears smaller than that of TMI (see Chapter 3), it seems likely that a major nuclear plant accident in the United States would set nuclear power development back for many years, if not permanently.

Yet another factor to consider is *radioactive waste disposal.* While the program to evaluate the Yucca Mountain site in Nevada appears to be making acceptable progress, major setbacks might well contribute to a social-political climate that would delay new reactor orders, even if demonstrably improved reactor technology is available.

Finally, dramatic *developments in non-nuclear electricity producing technology,* either positive or negative, could affect the attractiveness of new nuclear plants. For instance, should the evidence become clear that further fossil fuel burning will have catastrophic environmental impact, nuclear power development would become more likely. On the other hand, if new evidence or new technology substantially reduces environmental concerns about coal, substantial nuclear power development may become less likely.

The foregoing brief discussion of seven factors that could substantially affect new nuclear development is intended to set limiting conditions on the main scope of the present analysis. Development of nuclear reactors superior to those currently operating in the United States may prove to be a necessary condition for a revitalization of domestic nuclear power development. However, reactor technology development is not likely to be a sufficient condition by itself.

Again, the main goal of this analysis is to examine those factors that do appear to be related to reactor technology development. These include: 1) the financial climate for new nuclear power growth, and 2) the federal regulatory treatment likely to be displayed toward future reactor technology, 3) political behavior, which is closely related to public perceptions of safety, and 4) public attitudes, which have been heavily influenced by perceptions of safety. Some educated guesses will also be made about the future role of the media in influencing the public's perceptions of reactor technology developments.

After examining the apparent linkage between reactor technology and the financial climate, regulatory treatment, political behavior, and public attitudes, this study will conclude with an assessment of the prospects for new nuclear growth. Where possible, distinctions between the three reactor development paths will be drawn.

FINANCIAL, REGULATORY AND POLITICAL ISSUES

2

*T*he interplay of financial, regulatory and political factors in creating a set of conditions that affect nuclear power development was briefly introduced in Chapter 1. These interrelations are extensively discussed in the OTA (1984) report, Nuclear Power in an Age of Uncertainty.

The importance of financial, regulatory and political factors in enabling further nuclear development has been emphasized by the U.S. Department of Energy's Energy Research Advisory Board (1986). The Board urges, ". . . a comprehensive and integrated campaign to eliminate unwarranted institutional (primarily regulatory) impediments to the future development of civilian nuclear power." (p. 10, Vol. I)

Several distinctions should be made to define the scope of the current analysis. The first distinction is between the economics of nuclear power and the prospects for financing new nuclear power plant construction. The economic competitiveness of nuclear power has been extensively analyzed elsewhere (see for instance, OTA, 1984; Williams et al., 1987; USCEA, 1987; Critical Mass, 1987). These analyses focus primarily on a comparison of nuclear and coal as the two technologies available for large-scale, base-load power generation. We will not attempt to assess the various analyses or come to a judgment about whether, and under what conditions, nuclear power offers life-cycle cost advantages over coal. Projections of future costs for the two technologies depend heavily on a variety of assumptions. A partial list of these assumptions includes: future interest rates, rates of inflation, fuel costs, pattern of demand growth, regulatory changes, rate setting policies, technical development, and legislative actions.

The prospects for financing new nuclear power plants depend somewhat on the competitive economics of nuclear power. No utility would seek financing for a nuclear plant if there is, all things considered, a superior alternative at hand. However, life-cycle cost is but one of several factors utilities must consider in choosing among technologies.

The analysis here will be restricted to a consideration of the prospects for financing new nuclear plants. Where possible, attention will be given to the relation between prospects for financing and the three reactor technology development paths outlined in Chapter 1.

A second distinction relating to the scope of this analysis is between regulation at the federal level and the state level. Our analysis will focus primarily on the prospects for federal regulatory treatment of the three reactor technology development paths. The future behavior of PUCs is very difficult to predict and varies from state to state. There is also no clearly foreseeable relation between reactor development and likely treatment by PUCs.

A third distinction relates to the brief treatment we will give to the political environment. This will focus on the national political level rather than the state and local level. These latter political climates are undeniably important to individual siting actions. However, state and local politics are highly variable across locales and subject to rapid change as economic and environmental issues come and go. Local politics are not easily characterized from a national perspective.

Financing Issues

As a working assumption about the future economic viability of nuclear power, the conclusions of the 1987 analysis by a team from the Oak Ridge National Laboratory (Williams et al., 1987) will be accepted. They found that nuclear power would offer moderate life-cycle cost advantages over coal in all regions of the country except the north central region. These conclusions were based on a set of assumptions that included, ". . .a stable regulatory environment and improved planning and construction practices." (p. 1) The construction cost model represents the base construction costs for a small group of recently constructed plants whose costs are at the low end of the current range of costs. In other words, the study concluded that under a set of favorable (but not highly unrealistic) circumstances, nuclear power was likely to be cost competitive with coal for a plant starting up in the year 2000. The analysis was based on current-generation LWRs. This means, of course, that reactors developed by the three technology paths we are discussing (refined LWRs, advanced LWRs and advanced safety designs) must have life-cycle costs no greater than the projections for the "better than average" current generation LWRs.

The prospects for acceptance by the financial community of the reactors under development will be discussed in the context of the nine features of reactor technology development outlined in Chapter 1. Before undertaking this analysis, it is worth noting that if there is a key to acceptance by the financial community, it seems to be predictability. The amount of risk involved in an investment is, to a point, compensated for by expected return. However, since return on investment is limited by the rates set by PUCs, there is a constraint on the natural market mechanisms which would otherwise allow investment risk to be compensated. This means that uncertainties, particularly in construction and licensing, but also in operations, must be reduced by every means possible.

Plant Size   Smaller plant size appears to be one of the more important features in attracting financial backing. While economies of scale are still valid in theory, experience in the U.S. with constructing and operating the larger plants have, in most cases, not lived up to expectations over the last decade. The EPRI (1986) "requirements document" calls for reactors of smaller size. The DOE's Draft Program Plan for Advanced Liquid Metal Reactor Program (1987) also calls for smaller plant size, and DOE's program for developing advanced LWRs supports the development of both large and mid-size (approximately 600 MWe) plants.

The attractiveness of smaller reactors to the financial community is based on several factors. For one thing, the total investment in each reactor would be less, thus limiting risk. It is generally easier to get a small loan than a large one. For another, there is the perception that smaller reactors could be built more quickly than large ones. This would reduce the time period of the investment risk. Finally, small reactors can better match the curve of demand growth. This means construction can be started with more certainty that the power will be needed when construction is complete. Financial return in the early years of operation is thus more certain. Smaller reactor size is a key feature in the estimation of the financial community. It should be noted that the "small-is-better" design philosophy, current in the United States, is not universally shared. There is reportedly less emphasis on small reactor designs in Europe.

Inherent Safety   This feature by itself is not highly important to the financial community. The view is that any reactor likely to be constructed in the future will be safe enough. However, the potential for inherent safety to win public and political support is well recognized by financial people. Reactors with inherent safety features are in no way down graded—financial markets find new technology attractive if it is better technology, does not increase investment uncertainty and is no more expensive. Perhaps it is too early in the design process to project construction costs for the advanced safety reactors, but Weinberg and Spiewak (1984) estimate that the coming generation of American- British- and Japanese-designed LWRs will achieve a safety level of better than 10^{-5} core melt probability per reactor per year ". . .at the price of greater complexity and higher cost." (p. 1401) In other words, a hundred reactors of this type could operate with only one core melt expected in a thousand years. Contrary to the prediction that greater safety can only be achieved with a higher price tag, current development emphasizes design simplicity which should be both safer and less expensive to construct.

Operating Ease   In general, financing people are willing to "let the engineers worry about this." However, to the extent that greater operating ease and greater tolerance of operating transients serve to protect the equipment from damage, the financial community is highly interested. In the societal debate over nuclear power, attention has been focussed on preventing substantial radiation releases to the public. The TMI accident illustrates that even though radiation releases to the public were held to inconsequential levels during a severe reactor accident, the core was destroyed and the financial loss was very great. There are no plans at present to refurbish and restart TMI-2.

Weinberg and Spiewak (1984) estimate that the NRC's safety goal of 10^{-4} core melt probability per reactor per year (USNRC, 1983) has already been met or exceeded. They estimate further that the likelihood of a "substantial radioactive dose" to the public is 10 to 100 times lower than the likelihood that the reactor will be severely damaged. Since, in the eyes of the financial community, public safety seems adequately assured, severe damage to a reactor remains a more salient financial concern. Robust machines are thus attractive, whether achieved via inherent safety design or design for easier operation.

Construction Time   Shorter construction time is a very important feature to the financial community. As noted above, the duration of financial risk is shorter. There may also be advantages in dealing with PUCs, since there would be less opportunity for conditions and PUC personnel to change. The increased predictability—of need for power, of inflation, and of the social/political climate—is of great importance in evaluating investment risk.

Construction Cost   In borrowing money, an individual's ability to repay is generally more important than the absolute amount of the loan. Yet absolute amount is an important factor in securing nuclear power plant financing in the future. That is one reason why smaller plants are attractive. Construction cost per kWe is also highly important if nuclear plants are to be competitive with coal. The ORNL assessment (Williams et al., 1987) of nuclear power competitiveness with coal in the year 2000 assumed a construction cost of $1,690 per kWe for the reference case (better than average construction cost experience). This compared with $3,020 per kWe for the case more typical of recent construction experience. The OTA (1984) report describes as "reasonable" a construction cost goal of $1,300 per kWe (in 1982 dollars). This is based on a study of industry viability by the S.M. Stoller Corp. (1982), which estimates savings from fully standardized and pre-certified designs to be about 20 percent.

As noted previously, projected plant cost estimates are dominated by the assumptions one makes. Projections aside, reactor designers, vendors, utility management and AE firms should set a goal of construction cost "as low as reasonably achievable." Unless a good case for low cost and reduced construction uncertainty of future plants can be made, financing may be hard to attract. Robert Hildrith of Merrill Lynch (1987) has suggested that vendors, constructors and utilities form partnerships to ensure a common interest in keeping cost down.

It seems prudent to point out (OTA, 1984) that construction costs could also *increase* as a result of yet-to-be-recognized design and operational problems, which would likely prompt a new round of regulations.

Operating Cost   The financial community views operating cost as less important than construction cost. Operating cost during the early years of operation may be more important than later on. It is in the early years that construction debt must be retired. Operating efficiency as a result of efficient design makes a technology attractive to the financial community. Plants designed with ease of maintenance as a design criterion are also attractive. Yet the financial community recognizes that operating efficiency is heavily dependent on management effectiveness, a factor largely independent of technology.

Construction cost and operating cost, taken together, must enable whatever reactor technology is being considered to compete with supply alternatives, of which coal is the major competitor. Reactors that can operate efficiently at partial power levels and modify power levels quickly would offer additional advantages.

Modular Increments to Plant Size   The financial community did not show particular interest in nuclear plants built of small reactor modules. Yet looked at as a special case of the small reactor, which is attractive because of its ability to better match demand growth, incremental construction should be attractive. One concern is that the cost for construction of the non-reactor parts of a plant designed to accommodate a number of small reactors may be relatively high when only the first one or two small reactors are in operation.

Design Standardization   This feature is judged to be of very high importance because it affects so many other factors. It is viewed as a key to lowering construction cost, reducing the many uncertainties associated with constructing a one-of-a-kind design, and facilitating the regulatory process. These are all of great interest to the financial community. Standardized designs would also lend themselves to factory construction techniques.

As previously noted, institutional factors seem largely responsible for the minimal use of standardized designs in the U.S. There has also been a reluctance to "freeze" a design lest opportunities for technical innovations be lost. The financial community appears ready to lay such concerns aside. The issue of standardized design will remain somewhat academic until new orders are placed, but it would appear to offer many benefits if the institutional problems can be overcome.

Factory Construction   The attractiveness of this feature is dependent on its ability to lower construction cost, shorten the construction period and perhaps to ensure more uniform quality. Lower cost and quicker construction are important to attracting financing.

Several other conclusions have emerged from examining the prospects for financing nuclear power plants in the future. For one thing, the possibilities for the financial cooperation of vendors, constructors and utilities seem to be improving. Several such schemes have been considered. Some see better cooperation as a key factor in getting nuclear plant construction started again. Another factor which seems to be important is the management reputation of the utility. The past decade has provided many examples of the importance of effective management if costs and schedules are to be kept in bounds. For instance, the construction costs of 36 of 47 nuclear plants surveyed in 1983 by the Energy Information Adminstration were reported to be at least twice the initial cost projections. (*Time*, Feb. 13, 1984). There are many reasons for cost escalation, but to some observers poor management is an important one (Alexander, 1984).

On the question of "regulatory reform" the financial community, like the nuclear industry more broadly, remains convinced of the importance of a more timely and predictable licensing process. Yet there is a good deal of skepticism that rapid progress will be made. Expressions of good intentions aside, the financial community has seen few tangible signs of change.

On the question of whether a utility could raise financing today for a new plant, the answer seems to be, "with great difficulty." At this point, the faith in a positive construction cost scenario such as the reference case used by Williams et al., (1987) is just too low.

However, financial markets are notoriously changeable. Given the right set of circumstances, the picture could change within a few years.

Again, the question is premature since no utility is now known to be working actively to initiate a new nuclear plant project. An EPRI (1982) study showed that only one utility would consider a nuclear plant even if "generous financing rates" could be secured. However, other observers feel that some utilities would consider a nuclear plant "if they could afford one."

A general conclusion is that several more years will have to pass before the image of nuclear power can be restored sufficiently to permit serious planning for new reactors. In Hildreth's view (1987), evidence of better industry cooperation and a collective spirit of "we can do it" would help restore the image of the nuclear power industry in the eyes of the financial community. It goes without saying that during the process of rebuilding this image, plants must operate without incident. Bringing those plants still under construction into operation would also help restore confidence. This will not be a quick process.

The DOE's advisory committee on civilian nuclear reactor development (DOE, 1986) has called for a number of government initiatives, including a study of preserving the options to complete cancelled plants and initiatives in public information. The committee also recommended that,

> *"Studies should be initiated by the task force that will lead to recommendations for changes in state and federal laws and Federal Energy Regulatory Commission (FERC) administrative procedures to balance the financial risks of large central-station projects, both coal and nuclear, more equitably with the financial benefits." (p. 22)*

How these recommendations are acted upon and whether they succeed in creating a climate more conducive to financing new plants remains to be seen.

No firm conclusions can be drawn about how the financial community will react to the advanced safety reactor designs, beyond the generalization that new technology is favored if it is better. However, since the prospect of securing financing for such a reactor is still a number of years away, it is likely that financial community reactions to the advanced LWRs is more important in the near term.

The point was made earlier that predictability is the key to financing. Perhaps the three most important sources of uncertainty are construction cost, rate of demand growth and regulatory climate. While it may be too early to project the construction and operating costs of advanced LWRs and the advanced safety designs, there is a wealth of information on how construction costs escalate. Experienced and effective management and better industry cooperation should substantially reduce the uncertainty of construction costs.

Demand growth has obviously been difficult to predict in the past. As previously noted, the sharp and unpredicted reduction in demand growth in the early 1970s probably precipitated the wave of nuclear plant cancellations. On the other hand, demand growth in 1987 and 1988 is running more than double the predictions made only five years ago. Conservation measures in response to higher energy prices in the 1970s brought the demand growth curve down sharply, but given that conservation measures are subject to the laws of diminishing returns, it is increasingly less likely that major new reductions in energy use per capita will occur. The first new reactors to be ordered will surely be in response to demonstrable energy demand.

The regulatory environment is inherently difficult to predict because it reacts to social and political factors which can change markedly over a period of several years. Since new orders are not likely for several years, there will be opportunity for further observation of the regulatory environment. The next section will examine this environment in more detail.

Regulatory Issues

Assessment of the role of past federal regulatory actions in contributing to the decline of nuclear power is beyond the scope of this report. The OTA report (1984) devotes considerable space to an analysis of the impact of NRC actions on construction delays and cost. There is also the opinion expressed by many in the nuclear industry, that the large number of NRC requirements following the TMI accident, and especially the manner in which the NRC implemented new requirements, has added enormous and unnecessary cost, and may not have contributed to overall safety (NRC, 1981). The NRC survey (1981) of utilities reports,

> *"It is their [licensees'] perception that the cumulative affects of a 'cut/weld/fix' series of priority operations will ultimately end up hurting the plants." (p. 4)*

Since the focus of the current analysis is on reactor technology development rather than past regulatory action, the important question is how the NRC is likely to respond to the reactors being developed. There is an overall feeling in the nuclear industry that improvement or "reform" is needed in the way the NRC operates, if a more stable and predictable climate for nuclear power plant construction is to be realized. For instance, projections of lower plant construction cost in the future generally cite an improved regulatory climate as necessary to achieving reductions in construction cost and time, (OTA, 1984; Williams et al., 1987; USCEA, 1987).

Several issues bear importantly on the approach the NRC takes to the development of new reactor technology by industry and the DOE. These include: 1) the NRC's general attitude toward new reactor development, 2) whether new designs will be required to go through the full sequence of pilot test and demonstration, or whether they will be treated as evolutionary, 3) whether reactors with inherent safety features may receive regulatory safety credit, e.g., by having some systems now treated as reactor-safety-related removed from such treatment, 4) the willingness of NRC technical staff to work with reactor designers to help ensure the resulting design will be approved, and 5) whether the NRC transition from design-based to performance-based regulation will affect the regulatory treatment given advanced reactor designs.

Based on discussions with several observers close to the NRC, we can draw some tentative conclusions.

New Reactor Development   The NRC's Advisory Committee on Reactor Safety (1987) has included a dedicated decay heat removal system and greater design simplicity among the desirable features it would like to see in new reactors. The NRC staff is reported to be enthusiastic and cooperative in efforts directed at developing new reactors. At the same time, everyone recognizes it is not the NRC's role to "promote nuclear power." Nonetheless, the current level of cooperation is greater than it was some years ago. This is taken as a sign that the adversarial style that has often marked the NRC's approach is being tempered.

Regulation of New Designs   Since demonstration plants are expensive and would likely require scarce federal money and considerable time to construct, there is a good deal of interest in whether new designs can be treated as evolutionary and thus fall under extrapolations of existing regulations. The general view is that the advanced LWR designs might be treated as evolutionary since much of the technology has already been proven in practice, but that the inherent safety features of the LMR and HTGR would require step-by-step proof in order to win regulatory approval. Given the basic conservatism of the regulatory process, it does not seem wise to count on expedited treatment of new technology. Spinrad (1988), for instance, takes it for granted that,

> *"These advanced and alternative reactor types*
> *will need to have their safety, operability, and*

economy tested by experimental construction and operation." (p. 708)

He estimates that the time required by this process places the start of the "second nuclear era" around 2010 to 2015.

Credit for Inherent Safety   Weinberg and Spiewak (1984) have suggested the logic of regulatory credit for inherent safety using the Swedish-designed Process Inherent Ultimately Safe (PIUS) reactor as an example.

"Since there are no accident sequences that could release large amounts of radioactivity from PIUS, a high-pressure containment shell is unneeded." (p. 1402)

Earthquake proofing and elaborate emergency safety systems are also absent from the design. The attractiveness of down grading to non-safety-related, or omitting safety systems, is cost and simplicity. Weinberg and Spiewak feel that the economic success of the advanced safety reactors

". . . depends heavily on the commission's (NRC's) recognition that such reactors are fundamentally less prone to accident than current ones." (p. 1402) They wonder if the NRC ". . . will return to a less prescriptive mode of regulation." (p. 1402)

Yet, Haney and Blackman (1987) have identified the dominant risk for severe accidents (in LWRs) as human error. One technologist expressed his doubts in pithy terms, "Can we make *any* machine idiot proof?"

Observers of the NRC do not agree whether significant relief from current regulations is a realistic prospect. Since the loss of coolant accident (LOCA) is the current dominating concern, eliminating it should allow a number of regulations to be dropped. On the other hand, the NRC has been reported as quite reluctant to drop existing regulations in cases where new regulations would seem to make old ones unnecessary. Internal inertia, the conservatism of the regulatory process, and concern about criticism from nuclear opponents are the likely reasons. However, external criticism might be withstood if the safety case is a strong one and if a degree of social/political support for cost effectiveness can be secured. Such social support for cost effectiveness is likely to be hard to win. When there is the perception, whether correct or not, that public safety is involved, there is great reluctance to trade off safety and cost. However, people seem willing to make such choices in buying consumer products, e.g., deciding not to pay extra for optional anti-lock brakes on automobiles, or for installation of house smoke alarms. Developers of reactor technology will need to build an unassailable argument that some safety systems are completely unneeded.

NRC Cooperation on Reactor Development   This seems to be an area of genuine change that holds prospects for improved cooperation between the NRC and reactor designers. The current level of cooperation is viewed as encouraging, with a reduction of the historic rule of "arm's length." Yet, there will always be the traditional differences in role and perspective between developer and regulator and there will never be a prior guarantee that new technology will be accepted. As the development of advanced reactors moves beyond design and the time for regulatory action approaches, one might anticipate a move back toward traditional stances. To do otherwise could invite charges of improper collaboration, loss of perspective, and conflict of interest. Observers point out that NRC cooperativeness is less evident with LWRs. One remarked, "It's too late there."

One reason for pessimism about continued cooperation is the lesson of LWR regulation. Given that some problems do not become evident until during construction or operation, one can expect that a stream of problems will require regulatory rulings. History cautions that high expectations of rapid change are not realistic. Still, the current signs of some change toward greater cooperation is encouraging.

Performance Based Regulation   This trend in regulatory philosophy is seen to be more related to the regulation of operating plants than to the licensing process. Reactors designed and built with a "clean sheet" approach should theoretically outperform the existing models and would benefit commensurately from performance based regulation.

On balance, the prospects for some changes at the NRC in dealing with new technology are good. Given the many external constraints on the NRC's actions, the prospects for radical change are poor. The regulatory process is designed to be an open process that operates via public meetings. Decisions are made under the watchful eye of Congress, the technical community, the courts, influential critics, the nuclear industry and the media. Even if the structure of the agency is changed by one of the measures currently under consideration in Congress, it is doubtful if there will be the kind of relief from regulatory oversight that many promoters of nuclear power have called for. For one thing, the bulk of the survey data to be cited in Chapter 3 indicates that the public does not feel nuclear power has been over-regulated nor have the public's safety concerns eased. It would thus be difficult to generate the political pressure necessary to force change. New technology will have to prove itself as safe and the safety will somehow have to be achieved without economic penalty.

The analysis by OTA (1984) concludes that as long as there are "unsolved safety problems" with LWRs, there will be continuing regulatory pressure to correct those requiring correction. Thus, there is the prospect that even more backfitting may be necessary. Such problems provide powerful ammunition to nuclear critics in their arguments that the regulatory process not be subject to restrictions that could reduce its effectiveness. In such a climate, the current evidence of an open and receptive stance toward new reactor development is encouraging. The extent to which this open stance translates into continuing cooperation that results in a more predictable licensing process or less burdensome and costly regulatory action, remains to be seen.

Political Issues

As noted at the beginning of this chapter, the analysis of potential political actions with respect to new reactor technology will be brief for two quite different reasons: 1) politicians generally try not to reveal their positions on controversial issues, especially if public opinion is unsettled, and 2) analysis of public survey data appears to be an acceptable surrogate for political views. In other words, political actions are difficult to predict, partly because they tend to follow public opinion. The latter is quite uninformed on the specific subject of new reactor development.

Most nuclear industry supporters concede that congressional action on nuclear power issues has been historically pro-nuclear, despite feelings that much stronger governmental support is needed (DOE, 1986; The House Interior, 1986).

As of April 1988, six bills to restructure the NRC were under consideration. At least one of these (HR 4134) mentions standardized reactor design as a feature to be encouraged (*Nuclear Waste News*, 1988). Unless a politician's constituency is clearly pro- or anti-nuclear there is little reason to take a visible stand. Politicians have a duty to protect the health and safety of the public, and health and safety issues are frequently at the core of media attention to nuclear power. These factors taken together often leave politicians in the position of being quietly supportive of nuclear power development, but presenting a public stance of skepticism and vigilance.

The bind politicians face is illustrated by a survey by Cambridge Reports (Cambridge, 88/2). Asked if a candidate's position on nuclear energy would be an important consideration in their vote, 36 percent said it was not important. However, 31 percent said it was important and they would be more inclined to vote for a candidate who wants to continue developing nuclear energy. Twenty-one percent would favor a candidate who wants to shut down nuclear energy. Twelve percent were undecided. Politicians with constituencies like that of this survey would please ten percent more of the public by favoring than opposing nuclear development, but they would also risk being on the wrong side with the 48 percent who were not currently committed on this issue, but who might become committed as a result of events to come.

The distinction drawn below (see Chapter 3) between the involved public and the much larger segment of the public that is uninvolved in nuclear power issues, is relevant here. Politicians rank as involved by our definition, since most are relatively knowledgeable compared to most members of the public. They thus consider several nuclear related issues in addition to plant safety, e.g., need for power, cost, environmental impact, and waste disposal. The bulk of their constituency, however, is likely to focus on plant safety. Politicians must take a conservative stand on this issue regardless of their own personal views about plant safety.

All this leads to the tentative conclusion that political support for new reactor development and deployment will probably follow rather than lead public acceptance. This conclusion is in line with survey data comparing public attitudes to those of samples from 1) Congress, 2) federal regulators, 3) investors and lenders, and 4) top corporate executives. Based on data from Harris (80/1), Nealey et al., (1983) concluded that members of Congress expressed attitudes very highly similar to those of the general public on a variety of nuclear issues addressed by the Kemeny Commission after TMI. By contrast, the views of the other three "leadership groups" were substantially more pro-nuclear than those of the general public.

The same conclusion, that politicians' survey results are closely related to the public's and discrepant from other leadership groups, is supported by data from Harris (78/10) on the question of whether delays in nuclear power plant construction caused by public protests were a good thing or a bad thing.

On the question of the importance of nuclear energy in the years ahead, a 1987 Cambridge survey of 500 national opinion leaders (Cambridge, 1987) found 73 percent of federal legislators thought nuclear energy would be "very" or "somewhat" important. This compared to 82 percent for federal regulators, and 92 percent for financial executives. We report in Chapter 3 that the comparable figure for the general public in 1987 was about 78 percent.

The stance of politicians was less pro-nuclear than that of regulators and financial executives. The point is that the public will probably have to be won over before political support can be generated and that regulatory support is somewhat dependent on political support. Financing decisions are less directly influenced by political support, but must take it into account. A dramatic loss of political support could result in a situation such as that at the Shoreham Nuclear Power Plant. In remarking that Congress has shown "early signs of willingness" to consider supporting advanced reactor technologies, James J. O'Connor, (1988) CEO of Commonwealth Edison, also warned that,

> *"No utility in this country would consider ordering another nuclear plant without some assurance that, if we build the plant as approved, we will be able to operate it." (p. 4)*

The nature of the political process is such that it is difficult to say what form such an assurance could take. As previously noted, public acceptance must be won and kept in a variety of ways if nuclear power development is to resume.

PUBLIC ATTITUDES 3

*B*efore presenting data on public attitude trends it is necessary to ask the question, "What do public attitudes have to do with a resurgence of nuclear power?" After all, utilities are not in the habit of asking the public's permission to order a nuclear plant. Yet a good case can be made that without a certain level of public and political acceptance, utilities face an uncertain and uphill struggle in siting, financing and building nuclear plants, and in addition, securing from PUCs rates of return that justify the investment.

All survey data to be reported are from nationally representative samples unless otherwise identified. Surveys are identified in the text by a code which shows the year and month of the survey (e.g., "SA, 87/8" identifies a survey by Scientific American conducted in August 1987). Appendix A lists each survey code with the survey organization, sample size, and response mode.

The Office of Technology Assessment report (1984), describes six parties to the nuclear debate. All affect utility actions in various ways, and all are affected by public attitudes. The six parties are PUCs, investors, the NRC, nuclear critics, nuclear industry, and the public itself. The PUCs and the NRC are public bureaucracies that are much influenced by public and political support. Both use the public hearing as a formal mechanism for taking public views into account. Nuclear critics are dependent on public support for financial assistance and for a constituency. Critics without a constituency must rely on scientific credentials to make themselves heard. Yet, the overwhelming majority of credentialed scientists always have been and remain strongly supportive of nuclear power (SA, 87/8; CMPA, 87/4; Nealey et al., 1983; Rothman and Lichter, 1982). The nuclear industry and AEs are also somewhat dependent on public acceptance. The policy issues, including regulatory reform, on which the industry depends for its viability are in turn dependent on political and public support. Finally the public, expressing its will as ratepayers, through elected officials, and directly in state referenda, hearings and siting actions, constitutes a powerful force for or against nuclear power development.

Public opinion polls themselves influence elected officials, just as pre-election polls influence the behavior of political candidates. The media play an important role in both shaping and reporting public attitudes, which in turn affect political behavior and the actions of the various parties that influence utility decisions. The media tend to seek out and amplify controversy, particularly on public safety questions. Substantial media attention to an issue can create forces that politicians cannot ignore. This is not a charge of media bias, after all the media are under no obligation to balance "bad news" and "good news" on any topic (Nealey, 1979). However, it is a fact that the media have reported significantly more bad news than good news about nuclear power (Nealey, Rankin and Montano, 1978; Rankin and Nealey, 1979; Kemeny, 1980).

Considering the above points on the linkage of public attitudes to nuclear power development it is fair to ask the question, "What degree of public support might be necessary in order for nuclear power development to resume?" No quantitative answer is available, but it does seem useful to conceptualize the public influence on nuclear development in terms of "acceptance" rather than "favorability" or "support." Gene Pokorny, President of Cambridge Reports, Inc., a leading nuclear issues survey organization, has made this point persuasively (Pokorny, 1987). He argues that favorability toward nuclear power has declined in recent years but acceptance has increased. The author has cautioned elsewhere (Nealey and Radford, 1987) that nuclear power may have to settle for "grudging acceptance," much as most people accept going through airport security or visits to the dentist as necessary but not pleasurable activities. The point is that attitudes toward nuclear power involve many issues and intensities. If negative public attitudes achieve sufficient breadth and intensity, then nuclear power will not be able to develop. Somewhere short of this point, nuclear power may enjoy enough public acceptance to allow new plants to be constructed.

Opinions about the safety of nuclear power dominate the survey results and will receive major attention here. Much evidence points to safety as the key issue influencing the attitudes of the general public (Melber, Nealey, Hammersla, and Rankin, 1977; Nealey et al., 1983; Hohenemser, Kasperson and Kates, 1977).

A distinction should be made at this point between the "involved" and "uninvolved" segments of the public based on their degree of interest, knowledge, and involvement in the issue of nuclear power. While nuclear power is a rather mature social issue (having been debated for over thirty years), it is probably a major focus of attention for only a small minority of the general public. This is the case with other energy issues as well in the wake of the perception that the "energy crisis" is over. During the 1970s it was common for energy to be rated by up to 80 percent of the public as one of the *"most important issues facing the country"* (Pokorny, 1984). By January 1981 only five to seven percent felt energy was an important issue (Gallup, 81/1; YSW, 81/1). By June 1982 only one percent (YSW, 82/6) of the public rated energy a main issue. It has fallen off the charts since that time.

A reasonable estimate is that some 10-20 percent of the public is moderately interested and involved in the question of nuclear power. Ann Bisconti, Director of Research for the USCEA, places the figure at about 10 percent[a]. Combining survey data from 1979 and 1981, Miller and Prewitt (1982) report that about 17 percent of the public is interested and knowledgeable about energy policy. They refer to this group as the "attentive public." For this involved group, plant safety appears to be the major issue, at least among those opposing nuclear power. Other issues of medium importance to this group include need for power, cost, environmental issues and radioactive waste disposal. Among the remaining 80-90 percent of the public that is relatively uninvolved in the question of nuclear power, it is likely that plant safety is the only important issue, although perception of need for power would become important if they were faced with deciding on new plant construction.

Two conclusions follow from these generalizations. First, in reviewing the bearing that public attitude data have on the prospects for a resumption of nuclear power growth, perceived nuclear plant safety should be given prime consideration over other nuclear issues. This is especially the case for the current analysis which is focused on reactor technology development. In other words, most members of the public are likely to judge future reactors primarily on their perceived safety. Second, in reviewing the nuclear attitude data, one should keep in mind the fact that most respondents are not highly involved in or knowledgeable about the topic. This does not mean, however, that the attitudes of the uninvolved public should be ignored. In the event of a controversial nuclear plant siting (and all are controversial), a much larger segment of the local public becomes involved. Preexisting attitudes, even if not strongly held or well linked to the issues, will tend to influence the development of more strongly held attitudes.

(a) Bisconti, A. Personal Communication. May 11, 1988.

Two final points are germane to the interpretation of public attitude data. First, question wording has great impact on survey results. Nealey et al. (1983) give numerous examples of how apparently minor variations in the wording of nuclear attitude questions can substantially affect the results. For instance, in asking about the importance of nuclear power in the future, adding the phrase "to meet future energy needs" will significantly influence the results. The same would likely be true in surveys about the importance of "life insurance" versus the importance of "life insurance to take care of your loved ones if you should die."

Second, since nuclear attitude data consist almost entirely of the responses of individuals at one point in time, comparisons of similar questions at successive time intervals actually involve different samples, not the same individuals measured repeatedly. "Changes" in attitudes thus refer to changes in the percentages of samples that respond positively or negatively. Little is known about how the nuclear attitudes of individuals change over time, although Pokorny (1984) reported on one study of individuals polled in 1983 and again in 1984. He reported that 10 percent changed to a more positive position on nuclear power, 12 percent changed to a more negative position, and 30 percent and 43 percent were negative and positive, respectively, both times. The question wording was not reported.

Nealey et al. (1983) presented a large body of public survey data on attitudes toward nuclear power. That analysis included data through mid-1981, a point over two years after the TMI accident. The survey data to be reported here cover the period 1981 to early 1988, a point about two years after the Chernobyl accident. The TMI and Chernobyl accidents prompted a flurry of public attitude surveys for a period of several months in 1979 and again in 1986. The data to be reported here are organized around eight topics that bear on the prospects for further nuclear power development. These topics are: 1) public support for the general concept of nuclear power plant construction, 2) attitudes toward closing nuclear power plants, 3) attitudes toward building nuclear plants in the respondents' local area, 4) public views on the regulation of nuclear power, 5) public concern about nuclear power plant safety, 6) concerns about the Chernobyl accident, 7) beliefs and attitudes about nuclear power versus other energy producing technologies, and 8) beliefs about the need for nuclear power in the future.

Public Support for Nuclear Power Plant Construction

From the mid 1970s to the mid 1980s the classic measures of public support for nuclear power were two nearly identical survey questions used by Louis Harris and Associates and Cambridge Reports, Inc. The Harris question was as follows: *"In general, do you favor or oppose the building of more nuclear power plants in the United States?"* The Cambridge question was phrased, *"Do you favor or oppose the construction of more nuclear power plants?"* Nealey et al. (1983) provide complete data on trends in these two questions through Spring 1981. Harris found 56 percent favored, 26 percent opposed and 17 percent undecided in the period before TMI (April 1979). This changed to 49 percent favoring, 42 percent opposing and 9 percent undecided over the two years post-TMI. Cambridge data showed 51 percent favoring, 30 percent opposing and 19 percent undecided before TMI. In the two years after TMI, 45 percent favored, 40 percent opposed and 15 percent remained undecided. It appears TMI, and perhaps other factors such as declining need for additional electricity, reduced support by some 7 percentage points, increased opposition by about 12 to 13 percentage points and reduced the numbers of undecided respondents by about 6 percentage points.

The respondent's gender was substantially related to these differences before and after TMI. Favorability toward building nuclear power plants was about 10 percentage points higher for men than women prior to TMI. Comparing the pre- and post-TMI periods for men and women separately shows that favorability among men dropped by 2 to 3 percentage points, while favorability among women dropped by about 14 percentage points (Nealey et al., 1983).

In the current climate, when no new nuclear power plants are planned in the United States, these classic questions no longer provide much useful information. Data collected in May 1986 (ABC/WP, 86/5), reported only 19 percent favoring nuclear power plant construction, while 78 percent opposed. Only 3 percent were undecided. The question was phrased, *"In general, do you favor or oppose building more nuclear power plants at this time?"*

These data were collected only a few weeks after the Chernobyl accident. The media attention this accident received was very effective in bringing it to nearly everyone's attention. A survey (NBC/WSJ, 86/4) at the end of April 1986, found 92 percent of their general population respondents had heard or read something about the accident. Even allowing for the possibility that the question phrasing may have prompted positive responses, this is a surprisingly high figure.

Whatever the reasons—whether safety concerns in the wake of TMI, Chernobyl and other nuclear power plant mishaps, or because of a perception that more power is not currently needed—there is very little public support for building nuclear power plants *at this time.*

Yet this does not mean that at some time in the future, when a power shortage is perceived and perhaps when safety improvements become widely known, the public may not again favor nuclear power plant construction. In the same May 1986 survey (ABC/WP, 86/5) that found 19 percent favoring and 78 percent opposing nuclear construction, "at this time," respondents were asked, *"Would you describe yourself as a supporter of nuclear power plants as a means of providing electricity, an opponent of nuclear power plants, or haven't you made up your mind on nuclear power plants?"* Supporters totaled 27 percent, opponents 36 percent, while 35 percent hadn't made up their mind and 2 percent had no opinion. The relatively large percentage that hadn't made up their minds indicates the door remains open to support nuclear power development if the right conditions occur. On the other hand, presumably all respondents in this undecided category were currently opposed to building new plants. Events unfavorable to nuclear power could sway them into a more permanent negative position.

Additional evidence that opposition to building more plants is dependent on question wording comes from surveys in September 1985 and May 1986 (YSW, 85/9; YCS, 86/5). The question was, "In general, do you feel we should continue to build nuclear power plants or do you feel it's too dangerous to continue to build these plants?" The results were as follows:

	Continue to Build	Too Dangerous	Not Sure
YSW, September 1985	42%	49%	9%
YCS, May 1986	36%	57%	8%

Despite mention of danger, the support and opposition figures are much closer than the four to one opposition to building "at this time" noted in May 1986 (ABC/WP, 86/5).

These results are in line with those from a question with a complex introduction used by the Roper Organization in June 1983 and May 1986 (Roper, 83/6, 86/5). The question was, *"There have been a number of proposals for ways to reduce the risks of nuclear power plants. I'd like to ask you whether you favor or oppose each of them, bearing in mind that each one would either increase the cost or reduce the supply of electricity to some extent. Would you favor not permitting any more new nuclear power plants to be built, or don't you think the risks and costs justify that?"* The results were as follows:

	Oppose Not Building	Favor Not Building	Don't Know
Roper, June 1983	50%	35%	15%
Roper, May 1986	38%	52%	10%

While the question introduction raises safety risks as an issue, it also mentions cost and electricity supply. The effect of Chernobyl may account for the greater opposition to nuclear power construction in May 1986 compared to 1983 and 1985.

Attitudes Toward Closing Nuclear Power Plants

One finding that has remained remarkably constant is the overwhelming opposition to permanent closures of nuclear power plants. Nealey et al. (1983) report a variety of surveys in the months following TMI which showed only about 15 percent of the public in favor of permanently ending nuclear power in the United States. Roper (Roper, 83/6, 86/5), using the complex question introduction that refers to both safety risks and energy supply, asked in 1983 and again in 1986 about *"closing all existing nuclear power plants permanently."* Cambridge Reports in February 1987 asked about permanent closure of all 100 operating plants. The results from these three surveys follow:

	Oppose Permanent Closure	Favor Permanent Closure	Don't Know
Roper, June 1983	69%	17%	13%
Roper, May 1986	62%	28%	11%
Cambridge, February 1987	76%	20%	4%

The percentage favoring permanent closure did reach 28 percent just after Chernobyl, but a majority of better than 2 to 1 continued to oppose closure.

While permanent closure is seen as an extreme and unnecessary measure by most respondents, less drastic measures, if they are taken in the name of safety, are attractive to many. Nealey et al. (1983) provide data on seven surveys taken by Gallup, Harris and NBC News in 1979 and 1980. All pose *"cutting-back operations"* or *"closure"* until *"safety risks are better understood"* or until *"stricter regulations"* can be put in place. Just over 52 percent of those who stated their positions were in favor of these temporary restrictions in 1979 and 1980. About 10 percent had no opinion.

Public sentiment seems not to have changed very much on this point since 1980. Roper (Roper, 83/6, 86/5) used the same complex question that referred to cost and electricity supply reduction in surveys in June 1983 and May 1986 to pose the question, *"Would you favor closing all existing nuclear power plants until all equipment and procedures can be reviewed and improved, or don't you think the risks and costs justify that?"* The results were:

	Oppose Temporary Closure	Favor Temporary Closure	Don't Know
Roper, June 1983	43%	47%	11%
Roper, May 1986	40%	50%	10%

An intermediate proposal for curtailment of nuclear power was probed by two surveys by ABC News (ABC, 86/4) in April 1986 and by ABC News/*The Washington Post* (ABC/WP, 86/5) in May 1986. The questions proposed *"phasing out the nuclear power plants operating in the country."* The results showed the public opposed to such phasing out by a margin of about 55 percent to 40 percent with 5 percent undecided.

Another variation on this issue is the use of a three-part question that proposes 1) operating current plants and building more, 2) operating current plants but building no more, or 3) shutting down existing plants and building no more. Survey data from four time periods are available. The University of Michigan used this question in five surveys in 1980 (Nealey et al., 1983). The *Los Angeles Times* (LAT, 82/3) asked the question in March 1982. Cambridge Reports

(Cambridge, 84/Unk, 87/11) used this question in 1984 and in November 1987. The results show:

	Build More	Operate Only Existing Plants	Close Down	Don't Know
U. of Michigan, 1980	30%	45%	15%	10%
LA Times, 1982	28%	47%	20%	5%
Cambridge, 1984	30%	51%	17%	2%
Cambridge, 1987	24%	51%	18%	6%

Public sentiment has shown little change over the 1980-1987 time period.

Further indications that the public is reluctant to endorse more nuclear construction but is equally reluctant to forego the benefits of existing plants, comes from the various state initiatives and referenda since 1976. Nealey et al. (1983) analyzed such votes and attendant attitude surveys in Maine, Massachusetts, Missouri, Oregon, and South Dakota between 1976 and 1980. More recent votes in Maine (1982 and 1987), Missouri (1984), Oregon (1986), and Massachusetts and Nebraska (1988) support the earlier conclusions: 1) there is plurality or majority opposition to building new plants, 2) there is majority opposition to shutting down current plants, and 3) there is wide support for the proposition that the voters should be allowed to approve or disapprove proposals for nuclear construction. The analysis also showed that surveys taken prior to the elections were quite accurate in predicting the outcomes (Nealey et al., 1983). Such a finding lends credence to the predictive value of survey data on nuclear issues.

Attitudes Toward Building a Plant Locally

A general finding over the years since attitudes toward nuclear power were first measured is that sentiment toward nuclear power in

that people living close to nuclear power plants have more confidence in their safety than those who live far away (AP/MG, 86/6). However, in the wake of TMI and Chernobyl, any new plant siting in the United States is bound to be opposed by many local residents. As an indication of what nuclear plant siting faces, Roper (82/3) asked about siting a variety of facilities within 10 miles of the respondent's home. The respondents preferred, in order: a drug rehabilitation center, a mental hospital, a steel mill, a coal burning power plant, and a state prison to a nuclear power plant.

Remote siting may mitigate the NIMBY factor to some extent by placing fewer people in a zone where risks are perceived to be high. A survey of residents in the Harrisburg, Pennsylvania area four months after the TMI accident showed that the percentage of households in which anyone had considered moving dropped off rapidly with distance—from 30 percent within 5 miles to 19 percent within 5-10 miles, to 17 percent within 15-25 miles (Nealey et al., 1983).

Public Attitudes About Nuclear Plant Regulations

Virtually every observer of the decline in nuclear power development in the United States credits federal regulations as a significant factor (OTA, 1984; USDOE, 1986). Even the NRC's own survey (USNRC, 1981) of nuclear plant staff confirmed the widely held perception that the NRC should do a better job.

The question here is, what does the public think about the adequacy of regulations? We reported above that during 1979-1980 about 52 percent of those expressing a view favored restrictions or temporary closure of plants pending safety review or stricter regulations. Roper (81/8, 84/8), using slightly different questions in 1981 and 1984, asked whether there was too much, about right, or not enough government regulation of nuclear power. The results showed:

	Too Much	About Right	Not Enough	Don't Know
Roper, August 1981	9%	39%	43%	9%
Roper, August 1984	10%	30%	50%	9%

The *Los Angeles Times* (83/4) asked about the adequacy of *"federal laws and regulation about atomic energy plants"* in April 1983. Six percent felt they were "too strict," 29 percent felt they were "adequate," 43 percent thought they were "not strong enough," and 22 percent were not aware enough or didn't know how to answer. The Opinion Research Corporation (ORC, 84/9) in September 1984 asked about the level of confidence respondents had in *"federal government standards and procedures"* under which nuclear power plants were operating. Thirteen percent expressed "a great deal of confidence," 41 percent expressed "a fair amount of confidence," 28 percent had "little confidence" and 4 percent had no opinion.

The results of all these polls from 1979 to 1984 are highly consistent. The majority of the public apparently feels that regulations are either about right or not strict enough. Only about 10 percent of the public felt regulations had gone too far. In the face of these data, developing a public sentiment for reducing regulatory constraints on nuclear power appears unlikely. This is in line with the author's observation of a variety of surveys over the past ten years on public beliefs about a general easing of environmental regulations to permit industrial development. There is scant public support for such proposals. The public commitment to environmental protection and public health protection remains strong.

Of course "regulatory reform" as this term is used in the nuclear industry does not imply easing regulations or making any compromise with safety. Cambridge Reports (83/5), in a May 1983 survey of 1000 registered voters, probed directly a number of proposals for modifying the regulatory process and found wide support. The survey reported the public to be moderately well informed about the regulatory process and to be mildly critical of it, although the regulation of nuclear power was thought to be better than the regulation of other American industries. Regarding specific proposals for changes in the licensing and regulatory process, the survey found 75-90 percent of the voter sample to favor each of the following proposals: weighing plant changes against overall plant safety, weighing changes against cost to consumers, shortening the licensing period (without safety reduction), shortening the licensing period (if electricity bills decrease), earlier one-step licensing hearings, less formal hearings, pre-approval of standard designs, and early site approval. Without more information on the precise survey questions, it is difficult to say if these findings really give a green light to the proposed regulatory reforms or not. As discussed in Chapter 2, regulatory changes involve an extended political process during which public opinion develops and shifts. Proposals for regulatory changes must make it clear that absolutely no compromise with safety is involved. In fact, they should be able to make a demonstrable case for greater safety if they wish to garner sustained public support. Nuclear critics are sure to paint regulatory changes as dangerous relaxation of the controls on nuclear power.

Public Concern About the Safety of Nuclear Power

Early in this chapter, plant safety was identified as the issue dominating nuclear power attitudes. Nealey et al. (1983) report on a variety of studies in which various purported problems with nuclear power are compared. Based on data from Cambridge Reports gathered between 1975-1978, it is clear that perceptions of danger account for most of the reasons people volunteer for opposing nuclear power. A 1979 survey by the National Science Foundation (NSF, 79/10) found that *"meltdowns, nuclear explosions or another TMI type of accident, including human error"* led the list of harmful consequences that people expected from building more nuclear power plants. Surveys by Harris (Nealey et al., 1983) from 1975 to 1979 and by Associates for Research in Behavior (Nealey et al., 1983) from 1977 to 1980 make it clear that safety issues head the list of concerns about nuclear power.

Nealey et al. (1983) used regression analysis to statistically compare the reasons for opposing nuclear power plant construction. Safety was the most important factor.

Additional data come from a survey by Roper (82/7) in 1982. Reminded of news reports of accidents at nuclear power plants, 83 percent of respondents said they were "very" or "somewhat" concerned about such reports. Yankelovich, Skelly and White (83/6, 83/9, 83/12, 84/2, 84/12) probed several of the public's concerns about *"what's going on in the world,"* with the admonition that, *"you can't worry about everything all the time."* In five surveys between June 1983 and December 1984, between 73 and 79 percent of respondents said they worried "a lot" or "a little" about a nuclear power plant accident. Only 20 to 27 percent worried "not at all."

Curiously, the worry people express seems rarely to involve their own personal safety. Harris (83/8) and ABC News/*Washington Post* (83/4) found only 20 percent of their respondents indicated the area where they live was exposed to the threat of nuclear power plant accidents. In the same August 1983 survey, Harris (83/8) found that only 11 percent of the sample reported that nuclear power plant accidents had ever directly affected them or anyone personally close to them. Nealey and Radford (1987) have called opposition to nuclear power "altruistic": "I'm not threatened, but others are."

Whether or not people have direct experience with nuclear power, their feelings about permitting a new nuclear power plant in their area are based largely on perceptions of safety. These perceptions appear to come largely from media accounts, and as the Roper (1982) findings reported above show, they are concerned about what they see and hear in the media.

Two findings from a survey by Associated Press/Media General (86/6) in June 1986 are instructive. Asked if the nuclear industry had adequate safeguards to *"prevent a serious accident,"* 35 percent said "yes," 44 percent said "no," and 21 percent didn't know. Asked if the nuclear power plant closest to their home is safe enough or not, 41 percent said it was safe enough, while only 24 percent said it was not safe enough, and 35 percent didn't know. These respondents apparently were less concerned about their local plant's safety than safety of plants in general. This may result from the fact that most plants run with few incidents. Thus, people do not frequently hear of problems with their local plant. However, since TMI and Chernobyl, almost any plant incident, however minor, draws national media coverage. This means that the average person hears, with some frequency, about plant incidents, but these rarely involve the local plant. This underscores the importance of *all* nuclear plants achieving high reliability. The reputation of nuclear power is in the hands of each and every operating plant.

The public has little understanding of the actual operation of nuclear power plants. Operating incidents tend to be seen as signals that something is basically wrong. In other words, minor incidents have "signal value" far beyond their own importance (Slovic, 1987), much as frequent minor mechanical problems with one's car lead to the conclusion that it is basically flawed and a major breakdown is likely.

One classic indicator of knowledge about nuclear power operations is the belief, held by many, that nuclear power plants are subject to "massive nuclear explosions." Harris (75/4) measured this belief in April 1975 and found 39 percent who agreed, 24 percent who felt it was not possible and 37 percent who were not sure. In April 1979, just after TMI, Harris (79/4) found that 66 percent believed a massive nuclear explosion was possible. Twenty percent did not believe this was possible, and 14 percent were unsure. In yet another demonstration of the importance of the specific language in survey questions, a CBS (79/4) survey in April 1979 found only 36 percent of the public that believed *"a nuclear power plant accident could cause an atomic explosion with a mushroom shaped cloud like the one at Hiroshima."* Despite the differences in the

Harris and CBS findings, it is clear that many people harbor concerns about nuclear power plants blowing up. Chernobyl very likely reinforced this concern, although specific data to support this presumption are not available.

In line with the discussion in Chapter 1, that most people are not highly involved in or knowledgeable about nuclear power, the Associated Press/Media General (86/6) asked respondents if they thought *"most Americans know enough about the effects of radiation to make informed decisions concerning nuclear power?"* Only 16 percent said "yes, most Americans know enough"; 78 percent said "no"; 6 percent didn't know. Such a finding leaves us wondering if the respondents felt they themselves were knowledgeable enough to give informed opinions. In any case, it appears that most people realize the public is not well informed about nuclear power.

Once again, it is useful to remind ourselves that it is not necessary to be informed in order to hold an opinion or express an attitude, and that having an opinion will influence action, whether or not the opinion is informed or uninformed or based on knowledge or nonsense.

Data on the amount of risk people associate with nuclear power plants are available for several years. The results of five surveys from 1981 through 1986 are presented in Table 1.

It appears that less than half the public is currently satisfied with nuclear power plant safety. This figure appears to have changed little between 1981 and 1986.

However, in the survey by Associated Press/Media General (86/6) respondents were asked, *"Do you think that nuclear power plants in the U.S. are safer now than they were 10 years ago, or not?"* Sixty-three percent felt they were safer, 21 percent said they were no safer or about the same and 16 percent didn't know. Taking these findings at face value, we are left with the conclusion that despite the recognition that plant safety has improved, it is still not adequate in the view of many, if not most, Americans.

Table 1.   *Perceived Risks From Nuclear Power*

Survey	Date	
Roper (81/4)	April 1981	*Living near a nuclear power plant (How much risk is involved?)*

minor risk	moderate risk	high risk	don't know
22%	27%	46%	5%

Survey	Date	
Audits & Surveys (82/11)	Nov. 1982	*In your opinion, how safe or unsafe are nuclear plants in this country?*

very safe	somewhat safe	somewhat unsafe	very unsafe	don't know
14%	32%	24%	22%	8%

Survey	Date	
Harris (83/2)	Feb. 1983	*All in all, would you say that nuclear generating plants are mainly safe or do they tend to be dangerous?*

mainly safe	dangerous	not sure
41%	54%	5%

Survey	Date	
Opinion Research Corporation (84/9)	Sept. 1984	*How confident are you that most safety problems involving nuclear power have now been solved?*

very confident	somewhat confident	not very confident	not at all confident	no opinion
10%	37%	28%	22%	3%

Survey	Date	
AP/Media General (86/6)	June 1986	*In general, do you think the nuclear industry in the U.S. has adequate safeguards to prevent a serious accident from occuring, or not?*

has adequate safeguards	does not have adequate safeguards	don't know
35%	44%	21%

A survey by ABC News/*The Washington Post* (86/5) asked respondents to look into the future, *"In the future, do you think nuclear power plants can become a safe source of energy or not?"* Fifty percent said "yes" while 43 percent said "no, it cannot become a safe source." Seven percent didn't know. Despite the belief that safety has improved, it appears that many people have plant safety concerns that will be hard to satisfy.

This "ambivalence" toward nuclear power was also noted by Schneider (1986) after reviewing survey data since TMI, and is reflected in the results of a survey by CBS News (86/5). It asked, *"All in all, do you think the need for nuclear power outweighs the risk involved?"* The results were nearly evenly split. Forty-two percent said "yes" and 45 percent said "no." Thirteen percent didn't know. In political terms, the undecideds could swing the election.

While there is no doubt that perceived safety is the key to acceptance of nuclear power, it doesn't completely determine such acceptance. Public concerns must be interpreted against a backdrop of need for power, available alternatives and other factors. After all, we engage in many activities daily that involve known and recognized risks. It is not necessary for everyone to recognize the comparative safety of nuclear power in order for nuclear power growth to resume.

Attitudes About the Chernobyl Accident

The effects of the Three Mile Island 2 reactor accident in April 1979 on the attitudes of the American public have been exhaustively analyzed by Nealey et al. (1983). Clearly the TMI accident marked a turning point in public attitudes. Many who were positive about nuclear power before the accident had their confidence shaken. Others who hadn't paid much attention to nuclear power or were undecided took the accident as a signal that nuclear power was an unsafe technology. Those who had negative attitudes before the accident felt their views were vindicated.

The TMI accident did not have the catastrophic consequences (other than financially) so widely predicted by media accounts at the time. In retrospect, the accident can be interpreted to fit one's own prior beliefs. Those with preexisting negative attitudes about nuclear power could say, "Look what almost happened. Except for dumb luck there would have been a holocaust." On the other hand, many in the technical community took the accident as vindication of the robustness of reactor safety. They could say, "Despite a string of unfortunate technical problems and human error, the safety systems prevented all but insignificant radiation releases."

Following the TMI accident, polls showed the public did not feel the accident was a freak occurrence and that it could happen again (CBS, 79/4; Harris, 79/4). The CBS April 1979 survey also found that a majority of the public felt government officials had not been honest about the danger. Yet this survey found only 7 percent of the public felt the government was mostly to blame, while 15 percent blamed the utility. The majority, 55 percent, blamed human error, while 10 percent indicated several parties were at fault. Thirteen percent had no opinion.

Nealey et al. (1983) sum up the results in these terms,

> *"Although the general public, following the
> accident, did not favor shutting down all plants
> and still believed that nuclear power is important
> for meeting future energy needs, there was
> plurality sentiment for cutting back operations
> until stricter safety regulations could be put into
> effect." (p. 89)*

Additional regulations were put into effect and the industry took many other measures to increase safety. Apparently, this has had some impact on public attitudes. A survey in 1986 (AP/Media General, 86/6) found about two-thirds of the public felt nuclear power plants in the U.S. were safer than ten years ago.

This brings us to the accident at Chernobyl in April 1986. The media accounts of the accident were widely followed in the United States. As previously noted, an NBC News/*Wall Street Journal* survey in April 1986 (NBC/WSJ, 86/4) found that 92 percent of the sample had heard or read about the accident.

The accident also caused substantial concern about fallout in the U.S. In April 1986, an ABC News survey (ABC, 86/4) found 7 percent was "very worried" and 32 percent was "somewhat worried," *"that radiation from the Soviet Union nuclear plant accident will hurt you."* However, 61 percent was "not worried at all." In May 1986, a CBS News survey (CBS, 86/5) found 44 percent was *"concerned that nuclear fallout from the accident could affect you."* Fifty-two percent was not concerned. Four percent had no opinion. Gallup (Gallup, 86/5) found in May 1986 that only 29 percent felt there was a "serious risk" of fallout contamination in the U.S. Fifty-seven percent said there was not, while 14 percent didn't know.

U.S. public reaction was likely moderated by the fact that the accident was far away and happened in a country which many Americans feel is characterized by technological bungling. Still, most respondents to surveys at the time felt *"it could happen here."* An April 1986 survey (NBC/WSJ, 86/4) asked respondents to take into account the way *"nuclear power plants are built and regulated in the U.S."* Thirty percent felt it "very likely" and 40 percent said it was "somewhat likely" that such an accident could happen here. Only 29 percent said it was "not very likely"; 1 percent wasn't sure. Similar results came from a May 1986 CBS News survey (CBS, 86/5). The question mentioned the accident and asked if *"that kind of accident is likely to happen in the U.S."* The results showed that 55 percent thought it "likely" while 37 percent thought it "unlikely." A few respondents (1 percent) said "it already has" and 7 percent had no opinion.

Despite these pessimistic views, most Americans felt U.S. nuclear power plants are safer than those in the Soviet Union. Surveys by ABC

News in April 1986 (ABC, 86/4) and ABC News/*The Washington Post* in May 1986 (ABC/WP, 86/5), found 62 percent and 55 percent of respondents felt U.S. plants were safer than those in the Soviet Union. Two to three percent felt U.S. plants were less safe, but 26 and 35 percent of respondents to these two surveys felt the safety of plants in both countries was about equal. A few people (2 to 3 percent) volunteered that no plants are safe at all, while 6 to 7 percent had no opinion.

The long-term effects of the Chernobyl accident on nuclear power attitudes in the U.S. are difficult to predict. It is in the nature of the survey business that surveys cover currently hot topics. The Chernobyl accident prompted a flurry of nuclear attitude surveys. There were more such surveys in the spring of 1986 than for several previous years combined. As the event passed from public attention, surveys no longer covered it as a topic. Attitudes are also subject to a "rebound" effect. They tend to return toward the level that existed before an event affected them. In other words, the effects of TMI and Chernobyl tend to have substantial impact on attitudes for a while. If no similar events occur, attitudes tend to rebound toward pre-event levels. Nealey et al. (1983) documented the rebound effect in the months following TMI. We do not have sufficient data to document such an effect with Chernobyl, but it is likely that without more incidents and frequent reminders, attitudes will drift somewhat toward pre-Chernobyl levels.

It is also likely that Chernobyl, like TMI, has left a lasting legacy of doubt about nuclear power safety. A survey by ABC News (ABC, 86/4) found 58 percent of respondents were made "more fearful," by Chernobyl, of an accident happening in this country. Roper (Roper, 86/5) asked if the Chernobyl accident showed *"the inherent danger of nuclear power existing in all countries"* or only the weaknesses in nuclear systems and nuclear engineering in the Soviet Union. A majority (52 percent) felt Chernobyl showed the inherent danger in all countries. Thirty-four percent felt it showed only Soviet weaknesses. Two percent volunteered "both" and 11 percent didn't know.

A general opinion among observers of the nuclear power controversy in the U.S. seems to be that Chernobyl may have less long-term impact on public attitudes than did TMI, but that, given the much more serious nature of the Chernobyl accident, it has given nuclear critics fresh and powerful ammunition (Pokorny, 1987).

Nuclear Power Versus Other Energy Technologies

One of the requirements that nuclear power must meet if it is to experience renewed growth, is that it be superior, on balance, to the other energy supply alternatives. Of course, utilities do not ask the public to choose among energy technologies before making a decision. Nonetheless, public beliefs about the risks, costs and benefits of various technologies provide useful information. Such information helps identify the constituency nuclear critics can claim, it affords an opportunity to compare public beliefs with the opinions of energy experts, it gives an indication of the level of public knowledge, and it helps to set an agenda for public information efforts which will be necessary if the level of public knowledge is to be raised.

A series of nine survey questions asked by Roper (85/1) in January 1985 provides a comparative look at four energy supply technologies: nuclear power, coal, natural gas, and oil (see Table 2). The questions fall into three categories: safety and health, costs and benefits, and overall acceptability.

One caveat is necessary in interpreting these survey data: the question did not specify *electrical* energy sources. It simply referred to them as "possible energy sources."

The results provide some useful insights about the image of nuclear power. Despite the perception that nuclear power has a better track record than coal on past deaths, nuclear power is seen as by far the most dangerous energy source. There is little question that fear of a big accident dominates the public's image of nuclear power. In terms of costs and benefits, natural gas is seen as most environmentally benign and also most economical. Nuclear beats coal on environmental impact but coal is seen as more economical. Nuclear is seen to be in the best supply, although coal and natural gas are not far behind. Natural gas is a clear winner in acceptability for widespread use. Nuclear power is clearly the least acceptable.

It should be noted that surveys like this one that pit energy technologies against one another, tend to set up a popularity contest. Our earlier distinction between favorability and acceptability should be recalled. The point is that such comparisons may not be good predictors of how the public will respond to specific proposals to develop nuclear power in the future. Such proposals would necessarily involve cost, environmental, and availability analyses that will affect public opinion in the crucible of a siting controversy.

Nealey et al. (1983) provide extensive data and analyses on public opinions about energy alternatives up to mid 1981. Several differences, as well as similarities, from that period compared to the 1985 data are evident. In the former period, the health and safety implications of coal and nuclear power were seen as about equal. In 1985, coal had a decided advantage in the public's perception of health and safety. Nuclear power was seen as having less harmful environmental impact than coal in both periods and coal was seen as more economical than nuclear power in both periods. The potential supply of nuclear energy was judged to be a little better than coal in both periods and much better than oil.

One other difference from ten years ago involves the public's beliefs about the prospects for solar energy. A survey by Cambridge (75/4) in April 1975 found the public believed solar energy was cheaper than any other source including hydroelectric. The survey did not specifically ask about the perceived cost of *electrical* energy, so there could have been some confusion. A Gallup (79/4) poll in April 1979 showed that more people agreed (42 percent) than disagreed (36 percent) that solar-generated electricity could *"solve the energy crisis in the next five years.''* Public expectations of solar energy, sharply at variance with expert judgment, were typical 10-15 years ago. Perhaps that was a legacy of the "euphoric" treatment accorded solar energy by the media of the day (Rankin and Nealey, 1979). There are recent indications that public expectations of large-scale solar-electric energy contributions in the near future have decreased. A Cambridge (88/2) survey in February 1988 found only 14 percent of respondents believed solar energy would be the *"primary source of electricity ten years from now."* By comparison, 31 percent picked nuclear and 12 percent picked coal.

Table 2.   *Perceived Characteristics of Energy Supply Technologies (Source: Roper, January 1985)*

Safety and Health:
Source responsible for most *deaths through accidents over the past few years:*

Nuclear	Coal	Natural Gas	Oil	Don't Know
22%	33%	18%	7%	20%

Source responsible for fewest *deaths through accidents over the past few years:*

Nuclear	Coal	Natural Gas	Oil	Don't Know
21%	16%	23%	18%	22%

Source potentially the most *dangerous to human life:*

Nuclear	Coal	Natural Gas	Oil	Don't Know
82%	4%	5%	1%	8%

Source the safest *to use:*

Nuclear	Coal	Natural Gas	Oil	Don't Know
6%	26%	46%	16%	7%

Table 2. *Continued*

Cost and Benefits:
Source best *for the environment:*

Nuclear	Coal	Natural Gas	Oil	Don't Know
13%	8%	62%	8%	10%

Source most economical *for consumers:*

Nuclear	Coal	Natural Gas	Oil	Don't Know
14%	24%	44%	8%	11%

Source we're least likely to run out of:

Nuclear	Coal	Natural Gas	Oil	Don't Know
32%	27%	21%	9%	12%

Overall Acceptability:
Source most acceptable *for widespread use:*

Nuclear	Coal	Natural Gas	Oil	Don't Know
9%	9%	59%	16%	7%

Source least acceptable *for widespread use:*

Nuclear	Coal	Natural Gas	Oil	Don't Know
59%	20%	4%	9%	9%

Summing up the prospects, in the public's opinion, for nuclear power versus coal, it appears that nuclear power enjoys a slight advantage in supply and environmental impact. However, nuclear power is seen as a bit more expensive than coal and far more dangerous, despite the perception that nuclear power has caused fewer deaths than coal.

Beliefs About Future Need for Nuclear Power

The public's expectations about the need for nuclear power in the future may be some indication of public support in the future.

As background, it is evident that a plurality and probably a majority of Americans approve of the continued use of nuclear energy. Previously presented data on reluctance to shut down nuclear plants supports the contention of substantial majority support. Yet when asked directly about approving the use of nuclear power, the results are about even. Roper (86/5) asked about *"using nuclear energy to produce electric power."* The public approved 45 percent to 40 percent, with 15 percent unable to decide. On the other hand, CBS (86/5) asked if the *"need for nuclear power outweighs the risks."* Only 42 percent felt it did. Forty-five percent said it did not, and 13 percent had no opinion.

Cambridge (88/2) phrased a question as follows: *"Thinking about all energy sources available for large-scale use, would you say that nuclear energy is a good choice, a realistic choice, or a bad choice?"* This question is carefully worded to 1) get respondents to think comparatively about energy sources, 2) to consider availability, 3) to think in terms of large scale use, and 4) to be realistic rather than just expressing favorability versus unfavorability. The results showed 20 percent thought nuclear energy a "good choice," 49 percent a "realistic choice" and 27 percent a "bad choice." Four percent didn't know. Assuming that those who feel nuclear energy is a "realistic" choice would also "accept" it under the right circumstances, we have a substantial majority of 69 percent who appear to find nuclear power use acceptable. This is not the same thing as giving the go ahead to construct more nuclear power plants, although the use of the phrase *"available for large-scale use"* implies a future orientation.

The point has been made earlier that question wording has substantial influence on results. After all, a question is a verbal stimulus and the response it elicits varies as the stimulus varies. A question that mentions "energy need" or "to meet energy needs," or "if needed" will elicit higher approval of energy development than a question making no mention of need. Similarly, a question that refers to safety or risk or accidents or hazards will elicit lower acceptance.

Another influence on results comes from the survey context in which the question is embedded; in other words, from lingering reaction to previous questions in the survey. The location of the "good-realistic-bad choice" question in the survey is not reported. (Often question sequence is varied systematically from respondent to respondent to avoid biasing effects caused by the order of the questions.) However, other questions involving projections of nuclear power need in the future were included in the same survey. This by no means detracts from the usefulness of the findings but helps in their interpretation.

The same February 1988 survey by Cambridge (88/2) asked, *"How important do you think nuclear energy plants will be in meeting this nation's electricity needs in the years ahead?"* The results were: 53 percent "very important," 26 percent "somewhat important," 10 percent "not too important" and 7 percent "not at all important." Four percent didn't know. According to a press release by the U.S. Council for Energy Awareness (February 1988), this figure of perceived importance of nuclear power in the years ahead had increased from about 74 percent responding "very or somewhat important" during 1985 and 1986 to about 79 percent in 1987 and 1988. Moreover, Cambridge (88/2) also asked, *"Do you think the nation's need for nuclear energy as part of the total energy mix will increase in the years ahead or not?"* Seventy-six percent responded "yes," 17 percent said "no" and 7 percent were not sure. While these questions do not ask directly about the need for new nuclear power plant construction to meet future electricity needs, this is a plausible interpretation of the questions.

These findings are in line with other surveys of the same period. For instance, Gallup (87/9) also asked about the importance of nuclear energy plants in meeting electricity need in the future. The results were: 47 percent "very important," 30 percent "somewhat important," 13 percent "not too important," and 6 percent "not at all important." Four percent was undecided. A Cambridge (87/8) survey found nearly identical results.

The Cambridge (87/8) and Gallup (87/9) surveys also probed beliefs about the future use of generating technologies. Given a list of various sources used to generate electricity, respondents were asked to indicate *"Which one of these electricity sources do you think will be our primary source of electricity ten years from now?"* The results follow in Table 3.

While such data indicate that about a third of the public believe that nuclear power will be the primary source of electricity in 1997, they also demonstrate lack of understanding of the lead time necessary to build new plants, together with lack of understanding of the current energy supply mix. While these predictions for solar appear highly unrealistic, they are not as unrealistic as expectations for solar some years ago. A Roper (81/3) survey found 69 percent of respondents felt solar energy *"offers the best long-term source of energy in the year 2000."* This was followed by 34 percent who

Table 3. *Public Beliefs about the Primary Source of Electricity in Ten Years*

	Cambridge (87/8)	Gallup (87/9)
Nuclear Energy	36	31
Solar Energy	24	26
Coal	10	12
Hydroelectric	12	10
Natural Gas	7	8
Oil	6	8
Wind	2	2
Don't Know	3	4

indicated nuclear energy was the best long-term prospect. Coal was third with 28 percent. Of course, Roper's respondents in 1981 may have been expressing their wishes for solar energy rather than their expectations.

Summary

Public attitudes toward the construction of nuclear power plants were clearly positive until the spring of 1979. The TMI accident appears to have influenced attitudes toward plant construction to become more negative. Yet, a slight margin of support over opposition to plant construction continued until about 1982. At that time, support for continued nuclear power plant construction began to erode further. In 1987, sentiment was running four to one against new plant construction. As no new orders for nuclear power plants have been placed since 1978, the growth of public opposition can hardly be credited with stopping nuclear power. But, it is a clear impediment to a restart of nuclear power growth.

A combination of factors appeared to contribute to the decline of public support for nuclear power in the mid 1980s. The public felt the energy crisis was over; there were many problems with completing the plants that had been ordered; the public's long-standing belief that nuclear power is expensive seemed to be confirmed by the "rate shock" attendant to the start-up of each new plant; and a continuing series of plant reliability problems contributed to long-standing safety concerns. The Chernobyl accident probably contributed only a little to this slide in public support for nuclear power.

Building more plants is entirely different from the proposal, which arose after TMI, to shut down existing plants. In the wake of TMI, there was majority public support for proposals to curtail nuclear operations or for temporary closures to solve safety problems. However, only about 15-20 percent of the public has supported permanent closure. Barring major new events, there appears to be general is more positive than toward building a hypothetical plant in one's area (Nealey et al., 1983). There are several reasons. Risks posed by hazardous facilities tend to fall most heavily on neighbors. There is general reluctance to have industrial facilities of any kind sited near one's home. This general stance has been dubbed "not-in-my-backyard" (NIMBY). While recognizing that nuclear power plants may be necessary somewhere, many people would prefer the "somewhere" be elsewhere. Some respondents may indicate disapproval because they know there is no suitable site for a nuclear power plant in the immediate vicinity. After TMI, several surveys asked respondents if they would approve a plant *"within five miles of here."* Results from such a question are not of much use unless a suitable site exists near the respondent's home. However, comparative data over time do give an indication that local opposition has increased. Gallup (Gallup, 76/6, 79/4, 86/6) posed the question, *"As of today, how do you feel about the construction of a nuclear power plant in this area—that is, within five miles of here?"* Data from 1976, 1979 and 1986 are available:

	Don't Oppose	Oppose	Uncertain
Gallup, June 1976	42%	45%	13%
Gallup, April 1979	33%	60%	7%
Gallup, June 1986	23%	73%	4%

In line with the rise in opposition to plant construction in general, opposition to a local plant changed from about even to more than 3 to 1 against.

Such findings should not be taken as an indication that no plant can be sited anywhere. Future siting efforts will need to take local sentiment into account, as past siting efforts have done. There is also evidence that nuclear plants, once they are built, are often seen as "good neighbors" by residents in the area (Melber et al., 1977), and little prospect of public pressure forcing operating nuclear power plants to close down.

Support for nuclear power plant construction in one's local area has traditionally been lower than support for the construction of nuclear power plants "in general." The decline in support for local plant construction began several years before TMI while support for nuclear plant construction in general did not decline until TMI. Once constructed and operating, however, nuclear power plants often enjoy good local support.

Since TMI, a number of surveys have probed the public's views about regulation of nuclear power. The answer is unequivocal. The public favors very tight regulations. Proposals for regulatory reform are not likely to receive public support unless it is clear that they are for the purpose of increasing safety. Nuclear industry views that nuclear power has been over regulated seem not to be shared by the public.

Of the many reasons given for opposition to nuclear power, safety is paramount. As the most dramatic of the nuclear issues, safety commands attention even though most survey respondents do not pay close attention to nuclear issues. For many, safety is the most important, if not the only important, issue. There is no doubt that nuclear power presents an image of danger to many people. Half, or slightly more, of those responding characterize nuclear power plants as dangerous. Yet, few seem to feel personally threatened and a substantial majority of the public feels the safety of nuclear power has improved in the past ten years.

Media accounts of the Chernobyl accident were widely followed by the public and a source of widely felt, but not strongly felt, concern. Most recognized that U.S. reactors were different and safer, but felt that an accident of similar severity could happen in the U.S.

The public judges nuclear power to compare favorably with other generating technologies in two ways: it is not judged to have serious environmental consequences, and it is in nearly unlimited supply. Solar power is the sentimental favorite while natural gas is the most favored of the technologies that have been developed. Nuclear power is the least favored technology despite having recognized environmental benefits and a better safety record compared to coal. It is nuclear power's perceived *potential* for a catastrophic accident that seems to account for its lack of popularity compared to other technologies.

Despite strong reservations about the safety of nuclear power and very little support for nuclear power plant construction at this time, a large majority of the public continues to feel that nuclear power will play an important—perhaps increasingly important—role in supplying electricity in the future. Of the 75 percent or so of the public that feels nuclear power will be important in the future, most characterize it as a "realistic choice" rather than a "good choice." This underscores the reluctance with which nuclear power is accepted by most members of the public.

There continues to be a hard core of opposition to nuclear power that constitutes about one in four or five Americans. Ten percent or so of the public remains uncommitted, and the views, positive or negative, held by many members of the public are not strongly held views. Thus, many members of the public are open to the influence of future events and new information. The media are the primary means by which new events and information reach the public.

CONCLUSIONS 4

*I*n assessing the conditions that could lead to a resurgence of nuclear power development, it is necessary to start with the question of a perceived need for increasing energy supply. During the period of energy policy debates of the 1970s, it was common to think of nuclear power and solar energy as alternative energy choices. Amory Lovins' (1976) influential analysis of "hard-path" and "soft-path" energy strategies contributed to an easy over-simplification of the choices as nuclear vs. solar. Russell (1979) characterized the competing ideologies as "expansionist" and "limitationist." The latter contends that any imbalance in supply and demand of electricity should be addressed by conservation to limit demand. The former feels that an excess of demand over supply ought to be met by increasing supply. Dunlap and Olsen (1984) contrast activists from these two camps using survey data from Washington state. The two groups were about equal in their views of the seriousness of the energy situation, and were of about equal educational level. However, the hard-path activists were predominantly conservative and Republican, while soft-path activists were predominantly liberal and split between Democrat and Independent. There is a substantial literature on the characteristics and beliefs of supporters and opponents of nuclear power development. To go into it further would take us well beyond the scope of this analysis (see, for instance, Dunlap and Olsen, 1984, and Nealey et al., 1983) However, on the question of solar vs. nuclear, an analysis by Nealey et al. (1983), using survey data on acceptance of energy technologies, (ARB, 80/11) found that nuclear supporters were more accepting of all other energy technologies, including solar energy, than were nuclear opponents.

The point is that analysis of the conditions favorable to nuclear power development should start with the perception that there is a need for power which cannot or should not be met by restrictions on demand. Then, the question is how to increase supply. Only at this point does the issue of acceptance of nuclear power in comparison to other supply alternatives become relevant.

This point is important because it involves a limiting condition on the conclusion that plant safety is the most important issue in nuclear power acceptance. In other words, *if* there is a clear need for more power and nuclear energy is being considered, then plant safety appears to be the most important factor, at least to the general public.

Conclusions will be framed in terms of seven factors that affect acceptability of nuclear power to the five public sectors we have analyzed. Again, need for power is a background factor. The five public sectors are: 1) the uninvolved public, 2) the involved public, 3) the financial community, 4) the federal regulatory community, and 5) the political sector.

The seven factors that impact acceptance of nuclear power are: 1) nuclear plant safety, 2) cost, 3) non-nuclear supply technologies, 4) activities of opponents, 5) environmental impact of supply technologies, 6) perceived benefits of supply alternatives, and

7) radioactive waste disposal. Each will be briefly discussed. The conclusions on importance of the seven acceptance factors to the five public sectors are summarized in Table 4.

Nuclear Plant Safety   Perception of nuclear plant safety is dominated by the possibility of a catastrophic accident. Other safety issues include operating incidents that serve as signals to a skeptical public that nuclear technology involves unmanageable risks, and human factors that are seen as a major potential for accidents. The general public does not yet make distinctions among reactor technologies, but the regulatory sector does. The public may come to make such distinctions to a greater extent if there is the prospect of siting an advanced reactor. The mass media would figure prominently in shaping the perception of advanced reactor technology as different and safer than past reactors.

Cost   Construction cost and operating costs, including fuel cost, plant availability, and maintenance, are important. Construction cost appears more important than operating cost, but both must combine to contribute to competitive life-cycle cost.

Non-Nuclear Supply Technologies   Coal is the major alternative for large-scale base-load electric energy supply. Combustion turbines are important as an interim measure to meet peak demands, but if it becomes clear that a service area requires increased base-load capacity, coal is the likely competitive technology. Natural gas is also a consideration, but supply limitations and uses for natural gas other than base-load electricity production disqualify natural gas as a competitor to nuclear power in most areas. Technical advances in converting coal to electricity would, of course, increase its attractiveness. While not a supply alternative, major technical achievements in conservation, e.g., major energy savings as a result of applications of super-conducting technology, would reduce the need for new nuclear development as well as the development of other supply technologies.

In 1987, hydropower contributed about 10 percent of the electricity generated in the United States. However, according to estimates from Public Citizen's Critical Mass Energy Project (Rader, Bossong, Antypas, and Denman, 1988), hydropower generating capacity is expected to increase by only about 14 percent by the year 2000. The combined electric power contribution of the other renewables, including wind power, solar (both photo voltaics and solar thermal), biomass and geothermal, is estimated to be about 35 billion kWh per year at present (Rader et al. 1988). This amounts to about 8 percent of the electricity generated by nuclear reactors in the United States in 1987. Despite substantial increases in generating capacity from renewable sources in the last ten years, electricity from renewables appears too limited or too expensive to be a prominent competitor to coal for at least another decade.

Activities of Opponents   Anti-nuclear activists have mobilized opposition to nuclear plant siting in the past and can do so again if the issues on which they concentrate have broad appeal. Safety is the major issue although cost and need for power have been prominent issues in some siting controversies. The survey data show that some 20 percent of the public is strongly anti-nuclear. A much larger number, likely a plurality, is undecided or leaning, without much conviction, toward a pro- or anti-nuclear position. Many in this middle category will become more involved if a nuclear plant siting is considered in their area.

Environment Impact of Supply Technologies   Survey data as well as political action make it clear that the American public has a strong commitment to environmental protection. Reduced environmental impact is an advantage of nuclear power over coal that is widely recognized by the public. If additional evidence accumulates on the seriousness of acid rain or a build-up of carbon dioxide in the atmosphere, if the link between burning coal and these environmental problems becomes clear, and if the public comes to recognize that the problem is very serious, then coal might become significantly less attractive. If measures required to make coal burning environmentally acceptable are not feasible or become too costly, then coal would be less competitive. However, the dispersed effects of coal burning make it less likely that its environmental effects will be perceived by the public as catastrophic. In other words, concerns about the dispersed environmental effects of coal are unlikely to be as intense as the safety concerns of some area residents considering the siting of a nuclear plant in their area. Attitudes are much more likely to find expression in action when they are strongly held.

Yet, attention by the mass media, such as *Times'* recent article (Brand, 1988) quoting James Hansen of the National Atmospheric and Space Administration's Goddard Institute to the effect that the greenhouse effect has already begun, is the sort of coverage that affects public attitudes. The article concludes by noting that nuclear power is the main alternative to coal burning. A prominent *Newsweek* story in the July 11, 1988 issue (Begley, 1988) links the heatwave of summer 1988 to CO_2 emissions and notes that anti-nuclear groups are now admitting that nuclear energy is an option that should be considered. *Fortune* specifically covers new reactor technology and links nuclear power development to the greenhouse effect in an August 1, 1988 article (Faltermayer, 1988). Continued attention of this sort in the media and in the political sector could substantially change the odds that nuclear power will be judged superior to coal in at least some cases during the next decade.

Perceived Benefits of Supply Alternatives   Nuclear power is widely recognized by the public as being in nearly unlimited supply and domestically available. So is coal. Environmental impact appears to be the primary publicly recognized advantage of nuclear power over coal. All other technologies except oil are more popular with the public. If they are viable in terms of availability and cost, they will likely be preferred.

Radioactive Waste Disposal   Concerns about radioactive waste disposal have increased sharply over the past fifteen years. However, this issue appears unlikely to prevent a restart of nuclear power development unless it becomes clear that geological disposal is not viable. There is no question that many people feel intense concern over radioactive waste. For them, this issue is reason enough by itself to reject nuclear power. Continued progress of the waste disposal program will, over time, lessen the importance of this issue as a limitation on nuclear power growth.

Table 4 summarizes the apparent importance of the seven nuclear power acceptability factors to the five public sectors.

Table 4. *Importance of Factors Influencing Public Sectors*

	Public Sectors				
Acceptability Factors	Uninvolved Public	Involved Public	Financial	Regulatory	Political
1. Nuclear Plant Safety	high	high	medium	high	high
2. Cost	low	medium	high	low	medium
3. Non-Nuclear Supply Technologies	low	medium	medium	low	medium
4. Activities of Opponents	low	medium	medium	medium	high
5. Environmental Impact of Supply Technologies	low	medium	low	low	medium
6. Perceived Benefits of Supply Alternatives	low	low	low	low	low
7. Radioactive Waste Disposal	low	medium	low	low	medium

Before discussing the prospects for a resurgence of nuclear power in the context of Table 4, brief attention must be given to the role of the mass media in influencing the five public sectors.

Table 5 summarizes the direct and indirect effects of the media. A direct effect occurs when a sector gets its information almost entirely from the media. An indirect effect occurs when the direct effect of the media on one sector affects another sector. For example, the media have strong direct effect on both the uninvolved public and the involved public. A strong indirect effect on the political sector occurs because the political sector pays close attention to the way the public is influenced. This in turn generates moderate indirect effects on the financial and regulatory sectors. They do not depend on the media for information about nuclear power, but they may follow the media, as politicians do, as an indicator of public opinion.

Table 5. *The Strength of the Direct and Indirect Influence of the Media on Public Sectors*

	Public Sectors				
Media Influence	Uninvolved Public	Involved Public	Financial	Regulatory	Political
Direct	strong	strong	weak	weak	moderate
Indirect	weak	weak	moderate	moderate	strong

Prospects for a Resurgence of Nuclear Power

Financial Sector   The financial sector, as noted in Chapter 2, is interested in predictability. The major sources of unpredictability that were noted are need for power, construction cost, and regulatory climate. Of the acceptability factors in Table 4, only cost is rated as highly important. Plant safety is not a major concern of the financial community because there is confidence that proposed plants will be safe. However, the financial community is quite interested in the *perceived* safety of proposed technologies because of the potential for public opposition that generates difficulties in siting and licensing. Concern over the activities of opponents is related to their potential to influence the social, regulatory and political climate a new plant would face. At present, these climates are not positive enough or predictable enough to warrant financial community support for new plant construction.

Demonstrably improved reactor technology that could win better societal support would be welcomed, but improved reactor technology by itself seems not to be a requirement of the financial community. The financial community welcomes new developments in reactor technology that may result in nuclear power plants that will be cheaper to build and operate and less vulnerable to mishaps that cause plants to go out of service or result in major damage to the plant. As previously noted, a variety of factors, including improved management effectiveness and better cooperation within the nuclear power industry, could contribute to an improved image for nuclear power that would make financing new plants more likely.

Regulatory Sector   The analysis in Chapter 2 of the regulatory climate was restricted to assessment of the approach the NRC is likely to take to advanced reactor development. As Table 4 indicates, only nuclear plant safety receives a rating of high importance to the regulatory sector. Cost is very important to the regulators at the state level, but our analysis is focused at the NRC. The activities of opponents of nuclear power have medium impact on the NRC. The regulatory process is designed to respond when safety questions are raised from any quarter. Such response often means delay while the questions are examined and a determination is made about how to respond. The ability of anti-nuclear activists utilize litigation as well as to mobile social action involving thousands of demonstrators, letter writers and hearings attendees generates a degree of social/political pressure that the NRC cannot ignore.

We have noted the relatively open stance the NRC has taken to advanced reactor development. There is every reason to predict that the NRC would welcome continuing nuclear power development. After all, as a one-technology agency, its existence is tied exclusively to nuclear power. NRC staff appear, for the most part, to be willing to cooperate. There are some signs that the NRC has been stung by industry criticism. On an individual basis at least, a spirit of wanting to improve is evident. The ability of the organization as a whole to change its mode of operation, given the political constraints under which it operates, is less evident.

Political Sector   Table 4 indicates a pattern of importance of the seven factors affecting the acceptability of nuclear power to the political sector that is similar to that of the involved public.

The differences are that politicians give greater attention to the non-nuclear supply technologies and to the activities of opponents. Politicians must balance the interests of constituents, some of whom will be identified with other technologies. Of course the close attention paid to activities of opponents is a reflection of the social/political pressure they can apply, especially when a siting controversy flares up.

The survey data presented in Chapter 2 show that national politicians display nuclear attitudes quite similar to those of the general public and significantly less pro-nuclear than those of the regulatory and financial sector. It seems likely that winning the necessary degree of support from the broader public to enable a restart of nuclear power growth will also win the necessary political support. Increased scientific and media attention to the greenhouse effect during the heatwave of 1988 has already prompted Congressional attention to CO_2 emissions and increased support for nuclear reactor development. There is no reason to believe that political opposition to nuclear power will prevent its further development unless there is an increase in public opposition. As with the involved public, if there is a clear need for more base-load power, a nuclear technology that is superior in cost and environmental impact will stand a chance of winning political acceptance. A reactor technology that significantly reduces the probability of major accidents would enhance the chances of such acceptance, but only if the broader public also comes to such a conclusion. The media will play a significant role in this process.

Uninvolved Public   Currently, this sector does not play a strong part in contributing to or preventing conditions that could lead to new nuclear power development. Its primary importance lies in its potential to become more involved if new nuclear plant deployment becomes an active possibility. Survey data show the public (most of which is uninvolved by our definition) to be primarily concerned with nuclear plant safety. This sector opposes nuclear plant construction at this time but favors the continued operation of existing plants. It also apparently believes that nuclear power will continue to be important in the future and that this importance is likely to increase. Nuclear power is not a popular choice, but if more power is clearly needed and more acceptable alternatives are not available, this sector is likely to recognize that nuclear power is a realistic and acceptable choice. This scenario requires that there be no major new accidents and that the improving operation of existing plants continues to keep nuclear plant incidents out of the news.

While the uninvolved public is not cognizant of reactor designs, new reactor technology that is recognized by the expert community as substantially safer and is reported by the media to be substantially safer would help ensure acceptance. It seems unlikely that enhanced LWRs will capture such media attention. The smaller advanced LWRs and especially the new safety designs are more likely to do so. Of course, they must also be economically competitive to win the endorsement of the experts and thus have the chance to be considered by the general public.

Involved Public   This sector is also primarily concerned with nuclear plant safety, but as Table 4 indicates, several other acceptability factors are of moderate importance. This sector will evaluate the comparative economics of proposed nuclear plants, as well as other supply technologies and their environmental impacts. Progress in disposing of nuclear waste must continue. The activities of nuclear opponents are of some importance to this sector. Many will be active to some extent in groups that take a position for or against nuclear power. Developments in reactor technology are likely to be pictured by opponents as experimental, untried and risky. The involved public sector will evaluate these arguments using the media as a major source of information but also getting information from other sources. Public information programs can be an important source of information for this sector, whereas the uninvolved sector is likely to ignore them. During the controversies that will be generated by attempts to site new plants, the ratio of involved to uninvolved members of the public will increase. Members of the involved sector are more likely to have already formed opinions about nuclear power, but because they pay more attention to more issues than does the uninvolved public, a strong case for a superior technology stands a chance of winning acceptance.

This involved sector plays a strong role in the dynamics of public controversy. People in this sector are more politically active, attend public meetings, vote, belong to partisan groups, and talk to their neighbors and friends about the issues. The media pay attention to them. In short, their social influence is greater than their numbers.

Summary

Each of the five public sectors will have an impact on the prospects for a resumption of nuclear power development. In considering the set of conditions that would lead to a resumption of nuclear power development, five elements stand out.

First, there must be a clear need for more base-load power, together with clear indications that the need for power will continue to grow. Moreover, it must be clear that the imbalance of supply and demand for power cannot be mitigated by conservation measures. It must also be clear that other supply options that are currently more popular with the public will not be able to carry the load in the long run, or that they pose unacceptable environmental consequences.

Second, as the major available domestic alternative, coal will need to be perceived at least in some regions as less desirable than new nuclear power plants. To accomplish this, the next generation of nuclear reactors to be built must present clear life-cycle cost superiority over coal. Rate-setting policies by PUCs that help smooth out the front-end rate impact of nuclear power in comparison to coal would help nuclear power demonstrate life-cycle cost superiority and would also make financing easier.

Third, the reactors to be built in the future must feature substantially reduced cost of construction in comparison to most of the reactors constructed over the past ten years. This goal appears achievable based on the construction costs of some recent reactors. Greater use of standardized designs that would enable the licensing process to move quickly and smoothly would be important in reducing construction costs. Smaller plants may also be necessary. Securing financing would be easier, the construction period could hopefully be reduced and better matching of demand growth would be possible.

Fourth, the public must be convinced that a proposed nuclear power plant will be safe. Pokorny (1987) warns that safety will not be enough by itself. He points out that risks are never small enough unless there are compelling reasons to undertake them. Whether a dramatic technology advance will be necessary to influence the public's view of nuclear power safety is not known. Development and demonstration of the advanced safety reactors seems more likely to garner favorable media treatment than is the case with modifications of the LWRs, but this is an estimation made with limited confidence.

Finally, as noted previously, the nuclear power plants currently in operation need to show steady improvement in efficiency and enjoy a sustained period of quiet running.

REFERENCES

Alexander, C.P. (1984, February 13). Pulling the nuclear plug. *Time,* pp. 4-45.

Atomic Energy Clearing House. (1986). 32(25), 21-24.

Brand, D. (1988, July 4). Is the earth warming up? Yes, say scientists, but that may not explain this year's heat wave. *Time,* 132:18.

Begley, S. (1988, July 11). The endless summer? *Newsweek,* pp. 18-20.

Critical Mass. (1987). Too costly to continue. The economic feasibility of a nuclear phase-out. Washington, D.C.:*Public Citizen.*

Delene, J.G., Bowers, H.I., and Shapiro, B.H. (1988, October 30-November 3). *Economic potential for future water reactors.* American Nuclear Society/European Nuclear Society 1988 International Conference.

Denton, H.R. (1983). Nuclear power: Epilogue or problem? *The Energy Journal,* 4(1), 125-141.

Dunlap, R.E. and Olsen, ME. (1984). Hard-path versus soft-path advocates: A study of energy activists. *Policy Studies Journal,* 13, 413-428.

Electric Power Research Institute. (1982). *Summary of discussions with utilities and resulting conclusions: Preferred characteristic of new LWRs.* Palo Alto, CA:EPRI.

Electric Power Research Institute. (1986). *Advanced light water reactor utility requirements document.* Palo Alto, CA:EPRI.

Energy Information Administration, U.S. Department of Energy. (1986, March). *An analysis of nuclear power plant construction costs.* DOE/EIA-0485. Washington, D.C.

Faltermayer, E. (1988, August 1). Taking fear out of nuclear power. *Fortune,* pp. 105-114.

Fisher, C.F., Jr., Paik, S. and Schriber, W.R. (1986, July). *Power plant economy of scale and cost trends—Further analysis and review of emperical studies.* ORNL/SUB-85-7685/1&11. Oak Ridge National Laboratory, Oak Ridge, TN.

Haney, L.N. and Blackman, H.S. (1987). *Evaluation of human error estimation for nuclear power plants.* Idaho Falls, ID:EG&G.

Hildreth, R.G. (1987, November). A perspective on the financial community's views on nuclear energy. *Address to the U.S. Council for Energy Awareness,* Westin Century Plaza Hotel, Los Angeles, CA.

Hohenemser, C., Kasperson, R., and Kates, R. (1977). The distrust of nuclear power. *Science,* 196, 25-34.

The House Interior and Insular Affairs Subcommittee on Energy and the Environment received additional testimony on an array of issues concerning the status of the U.S. nuclear power industry. (1986). *Atomic Energy Clearing House,* 32(25), 21-27.

Kemeny, J. G. (1980, June/July). Saving American democracy: The lessons of Three Mile Island. *Technology Review,* pp. 65-75.

Lennox, F.H. and Mills, M.P. (1988). *Electricity from nuclear energy, burden or bargain?* Washington, D.C.:Science Concepts, Inc.

Lovins, A.B. (1976). *Energy strategy: The road not taken? Foreign Affairs,* 55, 65-96.

Melber, B.D., Nealey, S.M., Hammersla, J., and Rankin, W.L. (1977). *Nuclear power and the public: Analysis of collected survey research.* Seattle:Battelle Human Affairs Research Centers.

Miller, J.D. and Prewitt, K. (1982). *The American people and science policy.* New York: Pergamon Press.

Nealey, S. M., Rankin, W. L. and Montano, D. E. (1978). *A comparative analysis of print media coverage of nuclear power and coal issues.* Seattle:Battelle Human Affairs Research Centers.

Nealey, S. M. Perspective on public acceptance of nuclear power. (1979, December). *American Industrial Hygiene Association Journal,* pp. 1178-1190.

Nealey, S.M., Melber, B.D., and Rankin, W.L. (1983). *Public opinion and nuclear energy.* Cambridge, MA:D.C. Heath and Company.

Nealey, S. M. and Radford, L. M. (1987). Public fear of nuclear technology. *Waste Management '87.* The University of Arizona, pp. 171-178.

Nuclear Waste News. (1988). 8(17).

O'Connor, J.J. (1988). *INFO,* No. 231. Washington, D.C.:USCEA.

Office of Technology Assessment, U.S. Congress. (1981). *Nuclear power plant standardization.* Washington, D.C.:OTA.

Office of Technology Assessment, U.S. Congress. (1984). *Nuclear power in an age of uncertainty.* Washington, D.C.:OTA.

Pokorny, G. (1984, November). Address to the Atomic Industrial Forum, Hilton Hotel, Washington, D.C.

Pokorny, G. (1987, November). Address to the U.S. Council for Energy Awareness, Century Plaza Hotel, Los Angeles, CA.

Rader, N., Bossong, K., Antypas, A. and Denman, S. (1988, September). Turning down the heat. *Solutions to global warming: An analysis of energy efficiency, renewable resources, and other options versus new nuclear power development.* Washington, D.C.:Public Citizen, Critical Mass Energy Project.

Rankin, W. L. and Nealey, S. M. (1979). *A comparative analysis of network television news coverage of nuclear power, coal, and solar stories.* Seattle:Battelle Human Affairs Research Centers.

Rothman, S. and Lichter, S. R. (1982, August/September). The nuclear energy debate: Scientists, the media and the public. *Public Opinion,* pp. 47-52.

Russell, M. (1979). The process of making energy choices. In S. Churr, J. Darmstadter, W. Ramsay, H. Perry, and M. Russell (Eds.) *Energy in America's future: The choices before us.* Baltimore, MD:Johns Hopkins University Press.

Schneider, W. (1986, June). Public ambivalent on nuclear power. *National Journal,* pp. 1562-1563.

Slovic, P. (1987). Perception of risk. *Science,* Vol. 236, pp. 280-285.

S.M. Stoller Corporation. (1982). *Nuclear supply infrastructure viability study.* Argonne National Laboratory.

Spinrad, B.I. (1988). U.S. nuclear power in the next twenty years. *Science,* 239, 707-708.

U.S. Council for Energy Awareness. (1987). *A comparison of future costs of nuclear and coal-fired electricity: An update.* Washington, D.C.:USCEA.

U.S. Council for Energy Awareness. (1988, February). U.S. Public Opinion on Nuclear Energy (press release). Washington, D.C.:USCEA.

U.S. Council for Energy Awareness. (1988, March). *INFO,* No. 229. Washington, D.C.:USCEA.

U.S. Council for Energy Awareness. (1988, June). *INFO,* No. 232. Washington, D.C.: USCEA.

U.S. Department of Energy. (1986). *Review of the proposed strategic national plan for civilian nuclear reactor development.* Washington, D.C.:USDOE.

U.S. Department of Energy. (1987, March). *Nuclear energy long-range facility utilization plan (draft).* Washington, D.C.:USDOE.

U.S. Department of Energy. (1987, October). *Draft program plan, advanced liquid metal reactor program.* Washington, D.C.:USDOE.

U.S. Nuclear Regulatory Commision. (1981). *A survey by senior NRC management to obtain viewpoints on the safety impact of regulatory activities from representative utilities operating and constructing nuclear power plants.* Washington, D.C.:USNRC.

U.S. Nuclear Regulatory Commission. (1983). *Safety goals for nuclear power plant operation.* Washington, D.C.:USNRC.

U.S. Nuclear Regulatory Commission. (1987, January 15). *ACRS recommendations on improved safety for future light water reactor plant design.* Letter to Lando W. Zech, Jr. Washington, D.C.:USNRC.

Weinberg, A.M. and Spiewak, I. (1984). Inherently safe reactors and a second nuclear era. *Science,* 224, 1398-1402.

Williams, K.H., J.G. Deline, L.C. Fuller and H.I. Bowers. (1987). *Nuclear economics 2000.* Oak Ridge, TN.:Oak Ridge National Laboratory.

APPENDIX A

Survey Identification		Survey Organization	Approximate Sample Size	Response Mode
A&S	(82/11)	Audits & Surveys	Unknown	Telephone
ABC	(86/4)	ABC News	505	Telephone
ABC/WP	(83/4)	ABC News/*Washington Post*	1500	Personal
	(86/5)		1506	Telephone
AP/MG	(86/6)	Associated Press/Media General	1365	Telephone
ARB	(80/11)	Associates for Research in Behavior, Inc.	1000	Personal
Cambridge	(75/4)	Cambridge Reports, Inc.	Unknown	Unknown
	(83/5)		1000	Telephone
	(84/?)		Unknown	Unknown
	(87/2)		Unknown	Unknown
	(87/8)		1500	Unknown
	(87/11)		Unknown	Unknown
	(88/2)		1500	Unknown
CBS	(79/4)	CBS News	1158	Telephone
	(86/5)		695	Telephone
CMPA	(87/4)	Center for Media and Public Affairs	580	Unknown
Gallup	(76/6)	Gallup Opinion Index;	1500	Personal
	(79/4)	Gallup Poll	1500	Personal
	(81/1)		1609	Personal
	(86/5)		762	Telephone
	(86/6)		1004	Telephone
	(87/9)		1000	Unknown
Harris	(75/4)	ABC News-Harris Survey	Unknown	Telephone
	(79/4)		1000	Telephone
	(80/1)		Unknown	Telephone

Survey Identification		Survey Organization	Approximate Sample Size	Response Mode
	(83/2)		1253	Telephone
	(83/8)		1255	Telephone
LAT	(82/3)	*Los Angeles Times*	Unknown	Unknown
	(83/4)		1233	Telephone
NBC/WP	(83/4)	NBC News/*Washington Post*	Unknown	Unknown
NBC/WSJ	(86/4)	NBC News/*Wall Street Journal*	1599	Telephone
NSF	(79/10)	National Science Foundation	1635	Personal
ORC	(84/9)	Opinion Research Corporation	1019	Telephone
Roper	(81/3)	Roper Organization	2000	Personal
	(81/4)		1999	Personal
	(81/8)		2000	Personal
	(82/3)		2000	Personal
	(82/7)		2000	Personal
	(83/6)		2000	Personal
	(84/8)		2000	Personal
	(85/1)		1989	Personal
	(86/5)		1003	Telephone
SA	(87/8)	*Scientific American*	679	Unknown
YS&W	(81/1)	Yankelovich, Skelly &	1219	Telephone
	(82/6)	White, Inc.	1010	Telephone
	(83/6)		1007	Telephone
	(83/9)		1016	Telephone
	(83/12)		1000	Telephone
	(84/2)		1021	Telephone
	(84/12)		1024	Telephone
	(85/9)		1014	Telephone
	(86/5)		1000	Telephone
YCS	(86/5)	Yankelovich, Clancy, Shulman	1013	Telephone